Problem Books in Mathematics

Series Editor

Peter Winkler
Department of Mathematics
Dartmouth College
Hanover, NH
USA

Books in this series are devoted exclusively to problems - challenging, difficult, but accessible problems. They are intended to help at all levels - in college, in graduate school, and in the profession. Arthur Engels "Problem-Solving Strategies" is good for elementary students and Richard Guys "Unsolved Problems in Number Theory" is the classical advanced prototype. The series also features a number of successful titles that prepare students for problem-solving competitions.

Shyam Sunder Gupta

Creative Puzzles to Ignite Your Mind

 Springer

Shyam Sunder Gupta
Indian Railways
Jaipur, Rajasthan, India

ISSN 0941-3502   ISSN 2197-8506   (electronic)
Problem Books in Mathematics
ISBN 978-981-19-6567-8   ISBN 978-981-19-6565-4   (eBook)
https://doi.org/10.1007/978-981-19-6565-4

Mathematics Subject Classification: 00A08, 00A09, 00A07, 00A05, 97U40, 97N60

This Springer imprint is published by the registered company Springer Nature Singapore Pte Ltd.
The registered company address is: 152 Beach Road, #21-01/04 Gateway East, Singapore 189721,
Singapore

This book is dedicated with great respect to the memory of my mother late Smt. Shanti Devi, with the hope that it will keep alive her contribution in solving problems of all her near and dear ones.

Preface

The author takes great pleasure in presenting the book, *Creative Puzzles to Ignite Your Mind*, which is a great treasure for everybody who enjoys the beauty of the fascinating world of recreational mathematics. The book has been written keeping in view a large number of people who can be benefitted by reading the book. Apart from puzzle enthusiasts and math lovers, the book is considered of immense value for aspirants of Mathematics Olympiad, CAT/MBA and placement interviews of big companies like Google, Microsoft, Amazon, Apple, Facebook, Yahoo, NVidia, Oracle, Adobe, Morgan Stanley, Bloomberg and so on.

The book covers a large variety of intriguing puzzles with detailed ingenious solutions generally not found elsewhere. The puzzles can be browsed at random as these are not grouped in any orderly manner. The book is a result of collection and framing of puzzles to make them entertaining over a large period of thirty-five years, so references/sources of many puzzles are not available.

The book contains 25 short riddles and brainteasers in Chap. 1, which vary from simple but tricky to challenging ones. Chapter 2 provides solutions to all these short riddles and brainteasers given in Chap. 1.

Chapter 3 contains 125 creative puzzles of varying difficulty covering arithmetic and algebraic like clock, calendar, weight, age and digital puzzles, geometric puzzles, logical reasoning puzzles, combinatorial puzzles, match puzzles and game puzzles like who'll become a decamillionaire (kaun banega crorepati)?, new year winner, winning numbers, etc. Some famous and old puzzles like Cheryl's birthday, Bachets weight, liquid decanting, crossing bridge/river/desert, etc., have been included so that readers can find all types of puzzles at one place.

Only basic mathematics is required to solve these puzzles. But most of these puzzles are tricky and can be simplified by ingenious ideas. Chapter 4 provides detailed ingenious solutions to all 125 creative puzzles in a lucid manner along with special comments section which covers generalization of solutions, different techniques of attacking the same puzzle wherever possible. This will create interest in readers for further exploring the puzzle.

The title of puzzles has been suitably framed. Repetition of similar type of puzzles has been avoided to keep the book in concise form. However, important aspects of similar puzzles if any have been covered in comments section.

Like physical exercise is essential for a healthy body, mental exercise is equally essential for a healthy brain/mind. Puzzle solving is one of the brain exercises which can also train the mind to solve the real-world problems. So let us not waste time and start brain exercise as puzzles are eagerly waiting to ignite your mind.

Jaipur, India   Shyam Sunder Gupta

Acknowledgements

I sincerely acknowledge and thank Dr. Akhilendra Bhusan Gupta and Mukesh Singhal for their comments after going through some of the puzzles in the book. I thank my sons Tarun Kumar and Amit Gupta for going through the manuscript and helping in making all the illustrations in the book.

I thank my wife Sushil Gupta for her encouragement and support without which it would not have been possible to start writing and complete this book.

I got interested in puzzles after getting inspired by the puzzles of world's greatest puzzle makers like Sam Loyd, Henry Dudeney, Boris Kordemsky, Martin Gardner, Raymond Smullyan and others. I am grateful and thank them all.

Since the book has been prepared based on material prepared and collected over a long period of time, it is practically not possible to make available the references. However, I am thankful to everybody associated directly or indirectly with the puzzles in this book.

Every effort has been made to make the book error-free; however, some errors and mistakes may always remain. Therefore, I shall be grateful for any suggestions and comments not only for rectifying the errors and mistakes but for improving the book also.

Shyam Sunder Gupta

Contents

About the Author

Shyam Sunder Gupta is former Principal Chief Engineer, Indian Railways, Government of India. He is an Indian Railway Service of Engineers (IRSE) Officer of batch 1981. He has experience of more than 35 years in various managerial, administrative and technical positions such as Principal Chief Engineer, Executive Director, Divisional Railway Manager and Director/RDSO on Indian Railways. A recreational mathematician, who is actively involved in popularizing mathematics at the national and international levels, his major discoveries are equal product of reversible numbers (EPORNS), rare numbers, unique numbers, palindromic pseudoprimes, fifth-order prime polynomial, 17,350-digit memorable prime, etc.

His interest in number recreations dates back to 1978, when his first paper "Miracles of last digit" was published. Since then, his contributions have been published in *Science Reporter, Science Today, Math Education, The American Mathematical Monthly, The Mathematical Gazette, Mathematical Spectrum, Scientia Magna* and several other international books like *Unsolved Problems in Number Theory* (by R. K. Guy), *The Universal book of Mathematics* (by David J. Darling), *The Penguin Dictionary of Curious and Interesting Numbers* (by David Wells), *Prime Numbers: The Most Mysterious Figures in Math* (by David Wells), *Prime Curios!: The Dictionary of Prime Number Trivia* (by Chris K. Caldwell and G. L. Honaker) and *Those Fascinating Numbers* (by Jean-Marie De Koninck). He is co-author of the book, *Civil Engineering Through Objective Questions*.

List of Short Riddles and Brain Teasers

List of Creative Puzzles

Chapter 1
Short Riddles and Brain Teasers

1. **Health Monitoring**

   A survey was conducted in a town during the Covid-19 epidemic. It was found that 55% of the population is suffering from diabetes, 70% from heart disease and 80% from depression. Find the minimum possible percentage of the population suffering from all three diseases i.e. diabetes, heart disease and depression.

2. **The Surface Area and Volume of a Cube**

   The surface area and volume of a cube are both five-digit integers. The volume of the cube is an integer multiple of the surface area. Find the side of the cube. Assume the units of side, surface area and volume of the cube as mm, mm^2 and mm^3 respectively.

3. **Missing Numbers in the Sequence**

   Find the missing numbers X and Y in the following sequence:
   1100100, 10201, 1210, 400, 244, 202, X, Y, 100

4. **Weight of Rice Packets**

   There are five packets containing rice. Instead of weighing these five packets individually, they were weighed in all possible combinations of two packets i.e., packet 1 and 2, packet 1 and 3, packet 2 and 3 etc. The weights in kilograms of all these combinations in ascending order are 6.7, 7.2, 7.3, 7.8, 8.2, 8.6, 9.1, 9.2, 9.7 and 9.8. It is not known which weight pertains to which pair of packets. Find the weight of each packet.

5. **Solving Equations**

   Solve $(x^2 - 9x + 19)^{(x^2 - 15x + 56)} = 1$ and find all possible integer values of x.

© The Author(s), under exclusive license to Springer Nature Singapore Pte Ltd. 2023
S. S. Gupta, *Creative Puzzles to Ignite Your Mind*, Problem Books in Mathematics,
https://doi.org/10.1007/978-981-19-6565-4_1

6. **Pair of Balls**

A box containing 5 red, 11 green and 17 white balls are kept in a room. At least two balls of the same colour are to be taken out of the box, without looking at it. Find the minimum number of balls required to be taken out of the box in order to be sure of getting at least two balls of the same colour.

7. **Coin Denominations**

In a certain Kingdom, three gold coins of different denominations in tolas were in vogue. Anupama had four gold coins worth 20 tolas. Akshara had five gold coins worth 15 tolas. If both Anupama and Akshara had at least one gold coin of each denomination, find the value of the denomination (an integer number) of each of the three different gold coins.

8. **Buildings in the Colony**

The Buildings in a small colony are numbered using different positive integers. The sum of the building numbers is 101. If the maximum possible number of buildings exists in the colony, find the building numbers.

9. **Partially Completed Grid**

There are 16 squares in the grid shown below. Ten squares are filled with the letters A to J and six squares are filled with whole numbers. Replace the letters A to J in ten squares with numbers from 1 to 10, using each number only once, such that the six squares containing numbers become the product of numbers in their adjacent squares filled with letters.

			A				
		B	378	C			
	D	144	E	90	F		
G	160	H	40	I	60	J	

10. **Finding Birthday**

In 1940, a man stated that he was 2n years old in the year n^2. He also said that if his age in the year n^2 is added to the month number of his birth, the result obtained gives the square of the day of the month of his birth. Assume that n is a positive integer. Find the day, month and year of his birth.

11. **Allowance for Casting Vote**

During elections, it was decided by the administration to pay an allowance of Rs. 250 to every male and Rs. 200 to every female senior citizen who cast their votes. As per the census of the area, there were a total of 10000 senior citizens including both males and females. After the end of the voting day, it was noted that 60% of the male and 75% of the female senior citizens cast their votes. Find the total amount paid to senior citizens on the voting day.

12. **Counterfeit Cheque**

A person X bought an Antique Clock at a cost of $500 and sold it on the same day to person Y for $870. Person Y not having the change gave a $1000 cheque to person X. Person X not having cash at hand, requested another person Z, who was his friend for help. So, person Z paid in cash $130 to person Y in lieu of the $1000 cheque. Before paying the balance amount of $870 to person X, he (person Z) went to the bank to encash the $1000 cheque. However, to his surprise, the $1000 cheque was bounced and found to be counterfeit. So, person Z informed person X about the counterfeit cheque. Person X tried to contact Y but he was not traceable. So, person X repaid in cash $130 to person Z.
Find the amount actually lost by person X in the above transactions?

13. **Speed of an Old Car**

My son used to go to the school by cycle rickshaw. One day he left home at 7 AM but forgot his homework notebook. When I came to know about it, I started by my old car at 7:20 AM and after handing over the homework notebook to my son on the way, I returned home at 7:40 AM. Find the speed of my old car, if the speed of the cycle rickshaw is 20 km/h. Assume uniform speeds of cycle rickshaw and car in both directions. Ignore any time lost in handing over the notebook and turning back by my old car.

14. **Rate of Interest**

On a certain amount of money, the compound interest earned during the 6th and 7th year is 1000 rupees and 1060 rupees respectively. Find the annual rate of interest if compounding is done annually.

15. **Clock Digits**

A regular 12-hour digital clock displays hours, minutes and seconds like
08:37:56
At what time in each of the 12-hour periods, do all six digits of the clock change simultaneously.

16. **Unique Number**

Find the number N such that N^3 and N^4 together contain each of the ten digits i.e., 0, 1, 2, 3, 4, 5, 6, 7, 8 and 9 exactly once only.

17. **Largest Value and Average**

If the average of 20 different positive integers is 20, find the largest possible value that any one of the integers can have.

18. **Playing the Card Game**

Three persons namely A, B and C before starting to play the card game among themselves had balances as under:
A and B combined had five times as much money as C had, while B and C combined had four times as much money as A had. At the end of the play, their balances become as under:
A and B combined had four times as much money as C had, while B and C combined had three times as much money as A had.
If one person lost Rs. 100, find the amount each person had before starting the card game.

19. **Candle Light Study**

A student was preparing for his examinations. The light went off. The student lights two candles of equal length, one of which was thicker than the other candle. The thick candle was designed to last for five hours while the thin candle was designed to last for three hours. After completing his study, when the student went to sleep, the thick candle was three times as long as the thin candle. How long did the student study in the candlelight?

20. **Apparent Miscalculation**

This is an interesting ancient puzzle, where two fruit sellers used to sell apples every day in the market. One day, one fruit seller sold all his 30 apples at the rate of 3 apples per dollar, whereas, another fruit seller sold all his 30 apples at the rate of 2 apples per dollar. Thus, the total amount received by two fruit sellers is 25 dollars. It may be assumed that the apples of both fruit sellers are not distinguishable visibly, though the apples of the second fruit seller are of superior quality.
The next day, one fruit seller falls sick, so he requested the other fruit seller to sell his 30 apples also at yesterday's rate. Accordingly, the fruit seller went to the market to sell his 30 apples along with 30 apples from the other fruit seller at yesterday's rate. On the way, all apples got mixed up, so he decided to sell all the apples at the rate of 5 apples for 2 dollars. This appeared logical to him as three apples for one dollar plus two apples for one dollar are the same as five apples for two dollars. Thus, the total amount received is 24 dollars, one dollar less than yesterday.
How do you explain the difference of one dollar?

21. **Colliding Cars**

Two cars that are 60 km apart are travelling towards each other. The speed of cars is 108 km/h and 144 km/h. Find the distance between two cars just 1 second before they collide.

22. **Four Digit Squares**

If each digit of a four-digit perfect square is increased by 1, another perfect square is obtained, find all such four-digit perfect squares.

23. **Vanishing Squares**

Forty match sticks each 1 unit in length are placed in square form as given below:

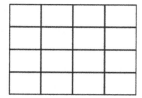

A total of 30 squares (1 of 4 × 4, 4 of 3 × 3, 9 of 2 × 2 and 16 of 1 × 1) can be seen in this arrangement. Find the minimum number of match sticks which on removing, vanishes all 30 squares.

24. **Two Different Date Formats**

Two different date formats i.e., DD/MM/YY and MM/DD/YY are prevalent in most of the countries. For example, 6th September 1960 is written as 6/9/60 in DD/MM/YY format and as 9/6/60 in MM/DD/YY format. Many such dates create confusion unless we know, which date format is used.
How many dates in a year can be written as two different valid dates in two different date formats which can create confusion?

25. **Product of Three Consecutive Even Numbers**

The product of three consecutive even numbers consists of seven digits. If the first and last digit of this product is 8 and 2 respectively, find the remaining five digits of this product and three consecutive even numbers.

Chapter 2
Solutions of Short Riddles and Brain Teasers

1. Health Monitoring

Let us first consider data for any two diseases, say diabetes and heart disease. 55% of the population is suffering from diabetes and 70% from heart disease. So, there is a minimum overlap of 25% as 55 + 70 − 100 = 25, which means at least 25% of the population is suffering from both diabetes and heart disease. Now consider two cases (i) % of the population suffering from diabetes and heart disease i.e., 25% as obtained above and (ii) % of the population suffering from depression i.e., 80% as given.

So, there is a minimum overlap of 5% as 25 + 80 − 100 = 5, which means at least, 5% of the population is suffering from all three diseases.

The same methodology can be applied for n diseases i.e., if percentages of people suffering from n diseases are P_1, P_2, P_3, ..., P_n respectively, then at least, P% of the population must be suffering from all n diseases, where P > 0 and is given by,

$$P\% = (P_1\% + P_2\% + P_3\% + \cdots P_n\%) - (n - 1) \times 100$$

2. The Surface Area and Volume of a Cube

Let the side of the cube is L mm. So, the surface area and volume of the cube are $6L^2$ and L^3 respectively. For the surface area to be of five digits,
$10000 \leq 6L^2 \leq 99999 \Rightarrow 40 < L < 130$
For the volume to be of five digits,
$10000 \leq L^3 \leq 99999 \Rightarrow 21 < L < 47$
Hence 40 < L < 47, so the side of the cube L (mm) can be 41, 42, 43, 44, 45 or 46.

Since the volume of the cube is an integer multiple of the surface area, so the side of the cube L must be divisible by 6, hence L = 42 mm. The surface area and volume of the cube are 10584 mm^2 and 74088 mm^3 respectively.

© The Author(s), under exclusive license to Springer Nature Singapore Pte Ltd. 2023
S. S. Gupta, *Creative Puzzles to Ignite Your Mind*, Problem Books in Mathematics,
https://doi.org/10.1007/978-981-19-6565-4_2

3. Missing Numbers in the Sequence

Each number in the given sequence represents the number 100 (base-10) expressed in bases $2, 3, 4, 5, 6, 7, 8, 9$ and 10.

So, the first number 1100100 in base 2 is equivalent to number 100 in base 10.

$$1100100_2 = 0 + 0 + 1 \times 2^2 + 0 + 0 + 1 \times 2^5 + 1 \times 2^6$$
$$= 100_{10}$$

Since $100_{10} = 144_8$, the number $X = 144$ in base 8 is equivalent to number 100 in base 10.

Similarly, $100_{10} = 121_9$, the number $Y = 121$ in base 9 is equivalent to number 100 in base 10.

So, the missing numbers X and Y in the given sequence are 144 and 121 respectively.

4. Weight of Rice Packets

Let the weight of the five packets from lightest to heaviest are denoted by A, B, C, D and E. It may be noted that A, B, C, D and E are all different because the weight of all combinations of two packets is different.

The smallest weight of the two packets is 6.7 kg, so,

$$A + B = 6.7 \tag{i}$$

The largest weight of the two packets is 9.8 kg, so,

$$D + E = 9.8 \tag{ii}$$

Obviously, $A + C = 7.2$ is the second smallest weight and $C + E = 9.7$ is the second largest weight.

It may be noted that the weight of each packet has been added to each of the four other packets, so each appears 4 times among the 10 combined weights of two packets each. So, $A + B + C + D + E = (6.7 + 7.2 + 7.3 + 7.8 + 8.2 + 8.6 + 9.1 + 9.2 + 9.7 + 9.8)/4 = 20.9$ kg.

From Eqs. (i) and (ii), $A + B + D + E = 6.7 + 9.8 = 16.5$. Therefore, $C = 20.9 - 16.5 = 4.4$ kg.

Since $A + C = 7.2$, so, $A = 7.2 - 4.4 = 2.8$ kg.

Since $C + E = 9.7$, so, $E = 9.7 - 4.4 = 5.3$ kg.

From Eqs. (i) and (ii), $B = 6.7 - 2.8 = 3.9$ kg and $D = 9.8 - 5.3 = 4.5$ kg.

So, the weight of the five packets of rice is 2.8 kg, 3.9 kg, 4.4 kg, 4.5 kg and 5.3 kg.

5. Solving Equations

To find all possible integer values of x to satisfy the given equation:

$$\left(x^2 - 9x + 19\right)^{(x^2 - 15x + 56)} = 1$$

Consider the following three cases,

(i) Zero-th power of any number (except 0) is always 1, so

$$x^2 - 15x + 56 = 0$$
$$\Rightarrow (x - 7)(x - 8) = 0$$

So x = 7 or 8. We can see that $(x^2 - 9x + 19)$ is not zero for x = 7 or 8.

(ii) Any power of 1 is always 1, so

$$x^2 - 9x + 19 = 1$$
$$\Rightarrow x^2 - 9x + 18 = 0$$
$$\Rightarrow (x - 3)(x - 6) = 0$$

Hence x = 3 or 6.

(iii) Any even power of -1 is always 1.

We can see that $x^2 - 15x + 56$ is always even for all values of x, so

$$x^2 - 9x + 19 = -1$$
$$\Rightarrow x^2 - 9x + 20 = 0$$
$$\Rightarrow (x - 4)(x - 5) = 0$$

Hence x = 4 or 5

So, all possible values of x are 3, 4, 5, 6, 7 and 8.

6. Pair of Balls

The key point to note is that the number of balls of different colours is given to create confusion, as there is no role of this data in solving the problem.

It is clear that if three balls are taken out then there is always a possibility (maybe small) of getting all the three balls of different colours but if four balls are taken out, then it is certain that at least two balls will be of the same colour as the balls are only of three colours.

So, if there are N coloured balls in the box, only N + 1 balls are required to be taken out to be certain of getting at least two balls of the same colour.

7. Coin Denominations

Let the value of denominations in tolas of three different gold coins be denoted by X, Y and Z. So, as per the given conditions, Anupama must have three gold coins of denominations X, Y, Z and one gold coin of any one of the denominations say X. So,

$$2X + Y + Z = 20 \qquad (i)$$

Akshara had five gold coins worth 15 tolas only, so, in addition to three coins X, Y and Z, she cannot have any additional coin as X because in that case $2X + Y + Z > 15$ from (i).

So, Akshara can either have (a) two gold coins of different denominations i.e., Y and Z, or (b) two gold coins of the same denomination say Y.

For case (a) we have

$$X + 2Y + 2Z = 15 \qquad (ii)$$

From Eqs. (i) and (ii), $3X = 25$, which gives non integer value of X hence not admissible.

Now consider the case (b), where we have two gold coins of the same denomination say Y, in addition to three coins X, Y and Z. So,

$$X + 3Y + Z = 15 \qquad (iii)$$

From Eq. (i), it is obvious that $X < 9$.

From Eqs. (i) and (iii), $X = 2Y + 5$, which gives $Y = 1$, $X = 7$. Therefore, from Eq. (i) or (iii), $Z = 5$.

Hence the value of the denomination (an integer number) of each of the three different gold coins is 7 tolas, 1 tola and 5 tolas.

8. **Buildings in the Colony**

   Since building numbers are different positive integers and the sum is fixed i.e., 101, so to maximize the number of buildings, we shall start building numbers from 1 onwards.

   The sum of the first n numbers is given by

   $$1 + 2 + 3 + \cdots + n = \frac{n(n+1)}{2}$$

   Since $\frac{n(n+1)}{2} \leq 101$, so the maximum value of $n = 13$.

   Now $\begin{aligned} &1 + 2 + 3 + 4 + \cdots + 13 = 91 \\ &1 + 2 + 3 + 4 + \cdots + 14 = 105 \end{aligned}$ and

   So the maximum number of buildings possible is 13.

   Since $1 + 2 + 3 + 4 + \cdots + 12 = 78$

   So first twelve buildings can be numbered from 1 to 12 and the 13th building can be numbered as $101 - 78 = 23$.

   So, the building numbers in the colony can be 1, 2, 3, 4, 5, 6, 7, 8, 9, 10, 11, 12 and 23 in addition to many other possible solutions.

9. **Partially Completed Grid**

This problem can be solved in many ways but we must keep the trials to a bare minimum. We can see that E is common to 378, 144, 90 and 40 so the value of E will be equal to either the factor of HCF or HCF itself of these four numbers. Now HCF of 378, 144, 90 and 40 is 2. So, the value of E = 2 or 1.

Similarly, we can observe that 8 is the divisor of 144, 160 and 40 only, so, H = 8.

Now the remaining two factors of 40 are 1 and 5 (as 40/8 = 1 × 5).

Since E can be 1 or 2, so E = 1 and I = 5

The remaining two factors of 160 can be 2 and 10 only.

(As 160/8 = 20 and 20 = 2 × 10 = 4 × 5 but 5 is already taken by I)

Since D cannot be 10 as 10 is not a factor of 144, so D = 2 and G = 10.

Remaining factor of 144 is obviously 9, as 144/(D × E × H) = 9, so B = 9.

Remaining two factors of 378 are 7 and 6 as 378/(B × E) = 42 and 42 = 7 × 6.

Since 7 is not a factor of 90 hence A = 7 and C = 6.

Remaining factor of 90 is 3 as 90/(C × E × I) = 3, So F = 3.

So Remaining last letter J = 4.

Hence the values of A, B, C, D, E, F, G, H, I and J are 7, 9, 6, 2, 1, 3, 10, 8, 5 and 4 respectively.

10. **Finding Birthday**

Since $n^2 \leq 1940$, so let $n^2 = 1940 \Rightarrow n = 44.045$. Considering n = 44 years, the man was 88 years old in the year $44^2 = 1936 < 1940$.

If we consider n = 43, the man would be 86 years old in the year $43^2 = 1849$ i.e., 86 + 91 = 187 years old in 1940, which does not seem to be logical.

Similarly, n = 45 also does not give a valid answer. So, n = 44 years and the man was born in the year 1936 − 88 = 1848.

Now $88 + M = d^2$, Where M is the month number of his birthday and d is the day of the month. Since $1 \leq M \leq 12$, M and d are integers, M = 12 and d = 10. So, the man was born on 10th December 1848.

11. **Allowance for Casting Vote**

At first look, it appears that data is insufficient to solve the puzzle, which is not so when you actually solve the puzzle.

Let the total number of male and female senior citizens be m and n. So, m + n = 10000.

Total amount paid to male senior citizens (Rs.) = 250 × 0.6 × m = 150m

Total amount paid to female senior citizens (Rs.) = 200 × 0.75 × n = 150n

Total amount paid to senior citizens (Rs.) = 150m + 150n = 150 (m + n) = 150 × 10000 = 1500000.

Since the average amount paid to male and female citizens is the same, it is not necessary to find the number of male and female senior citizens separately.

12. Counterfeit Cheque

The actual loss of person X = The cost of purchasing the antique Clock + $130 (Returned in cash) = $500 + $130 = $630.

If the profit of person X (i.e., $870 − $500 = $370) by selling the antique clock at $870 is also taken into consideration, then his loss is $630 + $370 = $1000, which is the value of counterfeit cheque. If the $1000 cheque had been legitimate, person X would not have lost any money including his profit.

13. Speed of an Old Car

Since the speed of the car is uniform in both directions so the time taken by the car in going is equal to half the total time of car drive = (7:40 − 7:20)/2 = 10 minutes.

So I met my son at 7:20 AM + 10 minutes = 7:30 AM

Distance travelled by my son from 7 AM to 7:30 AM (i.e., in half an hour with a speed of 20 km/h) = 10 km.

Since 10 km is also the distance covered by car in 10 minutes, so the speed of my old car = 60 km/h.

14. Rate of Interest

If r is the annual rate of interest, n is the number of years and S is the initial amount, then Compound interest (CI) is given by

CI = $S (1 + r/100)^n$, where compounding is done annually.

If we try to solve the problem using the above formula, things will get highly complex. The problem can be solved in a simple way as under:

Assuming that compounding is done annually,

CI for nth year = CI for (n − 1)th year + simple interest on the amount of CI for (n − 1)th year.

So CI for 7th year = CI for 6th year + simple interest on the amount of CI for 6th year.

That gives $1060 = 1000 + \frac{r}{100} \times 1000 \Rightarrow r = 6\%$

So, if compound interest is known for two consecutive years, the rate of interest can be easily calculated as given above.

15. Clock Digits

There are two digits each, used for displaying hours, minutes and seconds. Both digits will change when the last digit is 9 in the display of hour's digit, minute's digits and second's digits. So, we get display as X9:Y9:Z9.

Now X9 cannot be more than 12, so X = 0.

So, two digit hour display will be 09, which on changing will show 10 which means the minute hand display must be 59 and the second hand display shall also be 59. Hence Y = 5 and Z = 5.

So, the display will be **09:59:59**. At this time all six digits will change simultaneously to 10:00:00.

16. Unique Number

Since N^3 and N^4 together contain ten digits, so, the number of digits in N^3 and N^4 must be 4 and 6 respectively.

For N^3 to be of four digits,

$1000 \leq N^3 \leq 9999 \Rightarrow 10 \leq N \leq 21$

For N^4 to be of six digits,

$100000 \leq N^4 \leq 999999 \Rightarrow 17 < N \leq 31$

Hence $17 < N \leq 21$, so N can be 18, 19, 20 or 21.

For N = 20 or 21, N^3 and N^4 end in the same digit. Hence N can only be 18 or 19.

Since the sum of all digits is divisible by 9, so, the sum of N^3 and N^4 must also be divisible by 9. For $(N^3 + N^4) = N^3(1 + N)$ to be divisible by 9, either N^3 must be divisible by 9 or N + 1 must be divisible by 9. Hence N = 18. It can be seen that $18^3 = 5832$ and $18^4 = 104976$.

It will be interesting to find another unique number M such that M^2 and M^3 together contain each of the ten digits i.e., 0, 1, 2, 3, 4, 5, 6, 7, 8 and 9 exactly once only. The required number M is 69. It can be seen that $69^2 = 4761$ and $69^3 = 328509$.

17. Largest Value and Average

Let x be the largest possible value. For x to be the largest possible number, the other 19 numbers must be as small as possible. Since all numbers are different positive integers, it is obvious that these 19 other numbers must be 1, 2, 3 ... 19.

So, we have

$$\frac{1+2+3+\cdots+19+x}{20} = 20$$

$$\Rightarrow \frac{19 \times 20}{2} + x = 20 \times 20$$

$$\Rightarrow x = 210$$

In general, if the average of n different positive integers is n, then the largest possible number, say N is given by

$$\frac{1+2+3+\cdots+n-1+N}{n} = n$$

$$\Rightarrow \frac{n(n-1)}{2} + N = n \times n$$

$$\Rightarrow N = \frac{n(n+1)}{2}$$

So the largest possible value of one number is equal to the sum of numbers from 1 to n.

18. **Playing the Card Game**

Let the total amount in Rs. of all the three persons had initially been X.

So, C had X/6 before starting the card game and X/5 after the end of the play.

Gain of C = X/5 − X/6 = X/30.

Similarly, A had X/5 before starting the card game and X/4 after the end of the play.

Gain of A = X/4 − X/5 = X/20.

Total gain of C and A = X/30 + X/20 = X/12 = Loss of B.

So, loss of person B = X/12 = 100 ⇒ X = 1200.

So, the initial amount held in Rs. by person A is X/5 = 240, and by person, C is X/6 = 200.

The remaining amount in Rs. i.e., 1200 − 240 − 200 = 760 is the initial amount held by person B.

19. **Candle Light Study**

Let the initial length of candles = l cm

Duration of candle used for study = h hours.

Length of the thick candle after completing the study

$$= l - l \times \frac{h}{5}$$

Length of the thin candle after completing the study

$$= l - l \times \frac{h}{3}$$

After completing the study, the length of the thick candle is three times the length of the thin candle, so,

$$l - l \times \frac{h}{5} = 3 \times \left(l - l \times \frac{h}{3} \right)$$
$$\Rightarrow h = \frac{5}{2} \, hours$$

So, the student studied for 2.5 hours in candlelight.

20. **Apparent Miscalculation**

By taking 3 apples from the first seller and 2 apples from the second seller, a lot of 5 apples can be made which can be sold at the rate of 5 apples for 2 dollars. Continue making lots of 5 apples by taking 3 apples from the first seller and 2 apples from the second seller. It can be seen that after 10 lots of 5 apples each, no apple is left with the first seller and 10 apples (superior quality) are left with the second seller. Now, these 10 apples of superior quality have to be sold at the rate of 2 apples per dollar and not 5 apples for 2 dollars. This difference in cost i.e., 10/2 − 10/(5/2) = 1 dollar explains the difference.

21. **Colliding Cars**

The easiest way to solve such types of problems is to find the total distance which can be covered by two cars in one second.
Distance covered by first car in one second = 108000/3600 = 30 m
Distance covered by second car in one second = 144000/3600 = 40 m
So the total distance covered by two cars travelling in the opposite direction in one second.

$$\Rightarrow 30 + 40 = 70 \text{ m}$$

So the distance between two cars just 1 second before they collide = 70 m.
A still simpler way to solve this problem is using the relative speed of two cars i.e., 108 + 144 = 252 km/h. So in one second, it will cover 252000/3600 = 70 m.
Note that the initial distance given between two cars is not relevant for solving the problem.

22. **Four Digit Squares**

Let four digit perfect square = A^2.
After increasing each digit of A^2 by 1, another perfect square is obtained, let it be B^2,
So, $B^2 - A^2 = 1111$
$\Rightarrow (B + A)(B - A) = 1111 = 11 \times 101 = 1 \times 1111.$
If B + A = 1111, A and B both cannot be of four digits.
Since (B − A) < (B + A), So B + A = 101 and B − A = 11.
$\Rightarrow$ B = 56 and A = 45.
Which gives $A^2 = 2025$ and $B^2 = 3136$.
So, there is only one four digit square i.e. 2025, which gives another perfect square by increasing all its four digits by 1.

23. **Vanishing Squares**

One match stick from the outer periphery will be required to be removed to eliminate the only 4 × 4 square. The corner square locations shall be avoided as these will eliminate only four squares (i.e., 1 of 4 × 4, 1 of 3 × 3, 1 of 2 × 2 and 1 of 1 × 1). Any other location will eliminate 6 squares (i.e. 1 of 4 × 4, 2 of 3 × 3, 2 of 2 × 2 and 1 of 1 × 1).
The next aim is to break the perimeter of all inside squares. For this at least 8 match sticks are required. One match stick shall be removed from every two consecutive squares in such a way that the perimeter of all inside squares is broken. In this way, a minimum of 9 match sticks will be required to be removed to vanish all 30 squares. One solution is given below, though there are others also.
The match sticks marked a, b, c, d, e, f, g, h and i shown in Fig. 2.1 shall be removed to vanish all 30 squares as shown in Fig. 2.2 (dotted lines show removed match sticks).

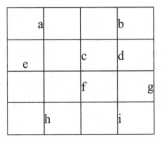

Fig. 2.1 .

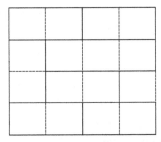

Fig. 2.2 .

24. Two Different Date Formats

There are twelve months in a year. In each month, twelve dates can be written differently in a valid form in two date formats. For example, 15th August 1947 is written as 15/8/47 in DD/MM/YY format and as 8/15/47 in MM/DD/YY format but 8/15/47 is not valid in DD/MM/YY (as MM cannot be more than 12). So it can be easily inferred that it is in MM/DD/YY format and cannot create confusion. Out of these twelve valid dates, one date, where DD and MM are equal like 8/8/47 will be the same in two different date formats. So, only eleven dates in one month can be written differently in valid form in two date formats.

Hence total number of dates in a year which can be written differently in valid form in two date formats = 11 × 12 = 132.

25. Product of Three Consecutive Even Numbers

Since the last digit of the product is 2, so the last digit of three consecutive even numbers will be 4, 6 and 8, as 4 × 6 × 8 = 192. The only other possibility is 2, 4 and 6, which gives 8 as the last digit and not 2 (as 2 × 4 × 6 = 48).

Since product consists of seven digits so the three consecutive even numbers must be three digits each. First digit of product of three consecutive even numbers is given as 8, so first digit of three consecutive even numbers must be 2 (as 2 × 2 × 2 = 8).

Since $(2.1)^3$ is more than 9, so first two digits of three consecutive even numbers must be 20. Hence three consecutive even numbers must be 204, 206 and 208 and product of these three numbers = 8740992.

Chapter 3
Creative Puzzles

1. **The Economist and a Mathematician**

On the first day of a New Century, the following conversation took place between an economist and a mathematician.

Economist says: If you give me one cent today, two cents on the second day (i.e., tomorrow), four cents on the third day, eight cents on the fourth day and so on i.e., doubling the amount each successive day till the end of the month, then in lieu of this, I can pay you 0.6 million dollars per day for the whole month. Would you like to avail this offer?

Mathematician replied: I may avail of your offer if you pay me 0.7 million dollars per day instead of 0.6 million dollars per day.

Economist says: I don't agree with your offer.

Mathematician says: Ok. I give you another offer, for which you can take time to think about and we can start the deal next month. As per this offer, I will ask for only 0.1 million dollars per day for the whole next month and I will pay what you asked initially i.e., 1 cent on the first day, two cents on the second day etc. for the whole next month.

Economist replied: I agree with your offer.

Can you find, who gained from the deal and how?

1 dollar = 100 cents.

2. **Mobile Phones and Pen Drives**

Two partners of a mobile phone shop decided to sell off all the stock of mobile phones and separately start a new business of selling pen drives.

Each mobile phone was sold for as many dollars as the number of mobile phones. From the sale proceeds of mobile phones, 64 GB pen drives were purchased at the rate of 10 dollars each. After purchasing 64 GB pen drives, the dollars left were less than 10, so they purchased one 32 GB pen drive from the leftover dollars (an integer number). Both the partners now have an equal

© The Author(s), under exclusive license to Springer Nature Singapore Pte Ltd. 2023
S. S. Gupta, *Creative Puzzles to Ignite Your Mind*, Problem Books in Mathematics,
https://doi.org/10.1007/978-981-19-6565-4_3

number of pen drives except for the fact that one partner received one 32 GB pen drive instead of a 64 GB pen drive. Find the cost of the 32 GB pen drive.

3. **Inter-hostel Athletics Competition**

During my college days, when I was staying in hostel no. 2, an Athletics competition was organized among three hostels i.e., hostel No. 1, hostel No. 2 and hostel No. 3. The following rules were to be observed.

Only one person from each hostel could participate in one event and participation was compulsory in each event from each hostel.

Points to be awarded to the first place, second place and third place are x, y and z respectively, where x, y, z are integers and $x > y > z > 0$.

I watched the Hammer throw event, in which the team belonging to my hostel got first place. Though the next event discus throw was started but I had to go for attending the class, so could not know the result of this event.

At the end of all events, hostel No. 1 won the overall championship with 24 points, and my hostel and hostel No. 3 both got 10 points each. Find the number of events held, the number of points awarded to first, second and third place in each event and the result of the discus throw event as to which team got first, second and third place.

Assume that there was no tie in any of the events.

4. **Ages and Inequalities**

A census surveyor asked a mathematician about the ages and number of his family members. The mathematician replied as follows:

"We are three in the family, me, my son and daughter. The sum of our ages is 100 years. My age is more than twice the age of my son. Five times the age of my son is more than six times the age of my daughter. Three times my age is less than eight times the age of my daughter."

All ages are in integer number of years.

Find the age of the mathematician, his son and his daughter.

5. **Consecutive Double-Digit Numbers**

Two consecutive numbers are selected from double-digit positive integers. One number, say X is given to person X and another number, say Y is given to person Y. Both persons are told openly that the numbers given to them are consecutive double-digit positive integers.

Can you find at least one number out of these two numbers based on the following conversation?

Person X: I don't know your number
Person Y: I don't know your number
Person X: Now I know your number
Person Y: Now I know your number

6. Interchangeable Hands of a Clock

Given a standard Analog two hands clock, when and how many times in a 12-hour period, the two hands of a clock i.e., hour hand and minute hand could be interchanged so as to yield another valid time.

For example, if it is 3 o'clock and the two hands of a clock are interchanged, the new position will not give a valid time.

7. Largest Product of Two Numbers

Find two numbers collectively using each of the digits 1, 2, 3, 4, 5, 6, 7, 8 and 9 only once, such that the product of the two numbers is the largest possible.

8. Three Balls and Four Boxes

Four identical boxes each containing three balls were labelled based on their content as WWW (Three White balls), BBB (Three Black Balls), BBW (Two Black and One White Ball) and BWW (One Black and Two White Balls). Later on, the labels of all the four boxes were disturbed such that none of the four labels correctly describe the content of the box.

Four persons were called and each was assigned one of the boxes whose label (i.e., incorrect) only he was allowed to see.

Each person was asked to randomly draw two balls from his assigned box and guess the colour of the remaining ball based on the label (incorrect) of his box. The first person after drawing two balls from his box declared that "He has drawn two black balls and he can tell the colour of the remaining ball in his box".

The second person after drawing two balls from his box declared that "He has drawn one white and one black ball and he can tell the colour of the remaining ball in his box".

The third person after drawing two balls from his box declared that "He has drawn two white balls but he cannot tell the colour of the remaining ball in his box".

The fourth person without looking at his box label and without drawing any ball from his box declared that "He can tell the colour of all three balls in his box and also the content of three boxes of other three persons too".

Can you find the logic based on which the fourth person was able to identify correctly the contents of all boxes?

9. Worm on a Rubber Rope

A worm is at one end of a rubber rope with an initial length of 1 m. The worm crawls along the rope towards the other end at a constant speed of 1cm per second. At the end of each second, the rubber rope instantly stretches by 1 m. Thus after 1 second, the worm has crawled 1 cm and the rubber rope becomes 2 m long after stretching. After 2 seconds, the worm crawled another 1 cm and the rubber rope becomes 3 m long after stretching and so on. Assume that rubber rope stretches uniformly, can be stretched indefinitely and the worm never dies.

Does the worm ever reach the other end of the rope?

10. **Winning Numbers for 101**

Two players A and B play a game. Let player A starts the game. He selects a number from 1 to 9 and asks player B to add a number from 1 to 9. To this sum, player A adds a number from 1 to 9 and so on. That is each player adds a number from 1 to 9 on his turn. The first one to get 101 is declared the winner. Find the numbers to be selected and the strategy to be adopted by the winner.

11. **Keys to Open the Locker Room**

The top four officials of a bank decided, on a system for opening the locker room of the bank as follows:

(i) No two officials can open the locker room by pooling their keys jointly.
(ii) Any three officials out of four can open the locker room by pooling their keys jointly.

Find the smallest number of locks required to be put in the locker room. Also find the smallest number of keys to the locks, which each of the four bank officials needs to carry to open the locker room as per the above system.

12. **Fastest Three Horses**

There are 25 horses and only five horses can be allowed to race at a time. Find the minimum number of races required to identify the fastest three horses, if no time measuring device is available. Assume that in each race, the performance of horses remains the same.

13. **Climbing the Stairs**

There is a single-storey building having a staircase with 10 steps. You can climb up either 1 or 2 steps at a time to go to the top. Find the number of distinct ways to climb the staircase.

14. **Shopping During Partial Lockdown**

I went to the market with a certain amount of money (in Rupees) in my pocket to purchase groceries, sweets, stationery, fruits and vegetables. For groceries, I paid an amount equal to half of the rupees in my pocket plus half a rupee. For sweets, I paid half of the remaining amount plus half a rupee. For stationery, I again paid half the remaining amount plus half a rupee. For fruits, I again paid half of the remaining amount plus half a rupee. For vegetables, I again paid half of the remaining amount plus half a rupee. After shopping, I went to the temple and donated half of the remaining amount plus half a rupee. While returning,

I purchased milk for the remaining amount of 100 rupees. How much amount in rupees did I was having initially when I went to the market?

15. Shaking Hands

I along with my wife invited five other married couples for dinner. As usual, some handshaking took place. However, no one shook their own hand or the hand of their spouse.

After the dinner, I asked each of the 11 other persons including my wife, how many hands, he (or she) had shaken. Surprisingly, each gave a different answer, so 11 different answers were received.

How many persons did my wife shake hands with?

Assume that no one shook hands with the same person more than once.

16. Equal Distribution by Doubling

Initially, the number of silver coins possessed by five persons namely A, B, C, D and E are a, b, c, d and e respectively, where $a > b > c > d > e$.

Now person A gives each other person i.e., B, C, D and E, the same number of silver coins as they already held that means, the person A doubles the number of silver coins of B, C, D and E. Then B did the same i.e., doubles the number of silver coins of A, C, D and E as they currently have. The same process is repeated by C, D and E.

It was found that finally each of the five persons held the same number of silver coins. Find the smallest number of silver coins each of the five persons possessed initially.

17. Identify the Apple Juice Bottle

A person buys seven bottles of juice in quantities of 10, 12, 16, 17, 21, 22 and 33 liters. Out of these, six bottles contain orange juice and one bottle contains apple juice. The person gives three bottles of orange juice to his neighbour and twice this quantity of orange juice to his relative. He kept the only remaining bottle containing apple juice with him.

Find the quantity of apple juice bottle.

18. Lockers Finally Remain Opened

There are 99 lockers in a school. The lockers are placed in a hall and are numbered from 1 to 99. Initially, all the lockers are closed. Now the following procedure is adopted to open/close the lockers.

  (i)   In the first step, all the lockers are opened.
  (ii) In the second step, every second locker (i.e., lockers numbering 2, 4, 6, 8 ...) are closed.
  (iii) In the third step, every third locker (i.e., lockers numbering 3, 6, 9, 12 ...) is opened if it is in a close position or closed if it is in an open position.

(iv) In the fourth step, every fourth locker is opened if it is in a close position or closed if it is in an open position.

(v) The above procedure is repeated till the last locker. So, in Nth step, every Nth locker is opened if it is in a close position or closed if it is in an open position.

How many lockers will finally remain open?

19. Weights in Increasing Order

Five silver items of different weights are to be arranged in increasing order of their weights. A balance scale without weights is available to compare the weight of the silver items.

Find the minimum number of weighings required to arrange the silver items in increasing order of their weights.

20. Grant Distribution Among Districts

The head of a state decided to distribute the grant among 25 covid affected districts i.e. $A_1, A_2, A_3, A_4 \ldots A_{25}$ as follows:

The first district A_1 gets 4% of the total grant.

The second district A_2 gets 8% of the remaining grant.

The third district A_3 gets 12% of the now remaining grant and so on i.e., the nth district gets 4n % of the available grant at his turn. The last district A_{25} gets 100% of the last remaining grant.

Find the district which gets the largest share of the grant.

21. Trucks to Transfer Payload

There are 5 trucks each with a full fuel tank of capacity 300 km. Find the maximum distance up to which the full payload can be delivered by using these 5 trucks if it is permitted to transfer fuel and payload from one truck to another anywhere on the route.

Assume that trucks can be driven with their own power only and empty truck can be dropped anywhere on the route. The payload of all trucks can be fitted in one truck if required. The mileage of trucks is independent of payload.

22. Inverting an Equilateral Triangle

Fifteen coins are placed in the form of an equilateral triangle as shown in Fig. 3.1.

Find the minimum number of coins required to be moved to invert the triangle (i.e., upside down) as shown in Fig. 3.2.

Fig. 3.1 .

Fig. 3.2 .

23. Largest Product from Distinct Numbers

Find the largest product which can be obtained from distinct positive integers whose sum is 100.

24. Gold Chain for Daily Payment

An Indian businessman went to London and rented a flat that required him to pay on daily basis. He was having a gold chain of 23 links. After negotiations with the flat owner, he agreed to give one link of the gold chain every day for 23 days stay. To minimize the number of cuts in the gold chain, it was also agreed that the flat owner will return some or all link/cut chains with him in exchange for a cut chain with one extra link for the day. For example, instead of giving one separate link every day, an Indian businessman could give one link on the first day, a cut chain of two links on the second day and take back one link given on the first day. So, every exchange was permissible subject to that every day flat owner must have total links equal to the number of days spent. Note that when a link is cut somewhere in between say 3rd link in the 7-link chain as shown in Fig. 3.3, three pieces of chain are obtained i.e., 1 link (where the cut has been made and can be restored by soldering), 2 link chain (i.e., left side of cut) and 4 link chain (i.e., right side of cut).

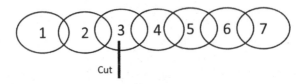

Cut

Fig. 3.3 .

    (i) Find the minimum number of cuts required in 23 link gold chain so that one link per day can be given as per the agreed conditions for 23 days stay.

   (ii) If an additional gold chain of 24 links is available with the Indian businessman and if he wants to stay for a total of 47 days, find the minimum number of cuts required in both the gold chains to enable him to stay for 47 days.

25. Age of Two Liars

Find the Age of two liars (whose statements are always false) based on the following conversation between them.

Liar A says, "My age is certainly not more than the number of playing cards in a deck i.e., 52."

Liar B says, "You are at least ten years older than me and I am forty-five years old."

Liar A says, "You are more than forty-five years old."

26. Three Sundays on Prime Numbered Dates

The birthday of my friend Om Prakash falls on the first day of a month in the second quarter of a calendar year. If three Sundays fall on prime numbered dates of the month of a year, find the weekday and month in which the birthday of Om Prakash will fall in that year.

27. Squares from Adjacent Numbers

Arrange the numbers from 1 to 15 in a row such that the sum of every two adjacent numbers is a perfect square.

28. Crossing the Bridge at Night

A family consisting of five members i.e., Grand Father, Father, Mother, Son and daughter wants to cross a narrow and shaky bridge, which can hold a maximum of two persons at a time. It was dark being night and they had only one torch which is necessarily required while crossing the bridge. Two persons can cross the bridge together at the speed of the slower person.

The time required to cross the bridge individually by Grandfather, Father, Mother, Son and daughter is 11, 6, 8, 1 and 2 minutes respectively.

(a) Find the minimum time required for the family to cross the bridge.

(b) If the daughter takes 5 minutes, instead of 2 minutes, what will be the time required for the family to cross the bridge.

29. Distinct Digit Milometer Reading

When I started my journey from Delhi to Chandigarh, I noted that my car milometer reading (in km) consists of five repeated digits. Find the shortest distance required to be travelled so that the car milometer reading (in km) consists of five different digits.

30. Three Hands of a Clock

At 12 o'clock, all the three hands of a clock i.e., hour hand, minute hand and second hand overlap. Find the times when these three hands of a clock will overlap in the next 12 hours.

31. Review Meeting Days

On 1st January 2020, the Managing Director of a company, after joint discussions with directors of all five departments namely operation, finance, design, business development and HR finalized the timetable for review meetings and conveyed as under:

From today onwards, review meeting with HR department will be on every alternate day i.e., 3rd January, 5th January, 7th January and so on. Review meeting with the operations department will be on every third day, with the business development department on every fourth day, with the design department on every fifth day and with the finance department on every sixth day.

(i) Find the day when the managing director will again meet with the directors of all the five departments jointly.

(ii) Before the next joint meeting with all departments, find the days when there is no review meeting with any of the departments.

Assume that review meetings will be held on scheduled days irrespective of holidays.

32. Curious Olympic Rings

The five Olympic rings overlap to give nine separate regions A, B, C, D, E, F, G, H and I as shown in Fig. 3.4.

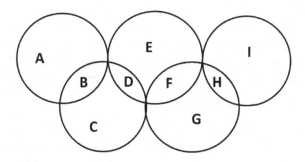

Fig. 3.4 .

Assign the numbers 1, 2, 3, 4, 5, 6, 7, 8 and 9 in the nine regions A to I such that the sum of the numbers in each ring is 14.

33. **Maximum Area of Plot**

There is a plot of land with four sides measuring 5 m, 9 m, 13 m and 15 m respectively. The area of this plot is 80 sqm. The owner of the land offered this plot for sale at a total cost of four million rupees. During the discussion with the purchaser, the owner gave the second option to the purchaser as follows:
"The purchaser is free to take as much land area for the plot as he is able to enclose with the given four sides i.e., 5 m, 9 m, 13 m and 15 m. However, for this option, the purchaser has to pay 10% extra i.e., 4.4 million rupees."
Can you help the purchaser to decide by finding the possible area of plot in the second option?

34. **Building Location for a Senior Citizen**

The sixteen residential buildings i.e. A, B, C, D, E, F, G, H, a, b, c, d, e, f, g and h are located along the road YZ in a colony as shown in Fig. 3.5.

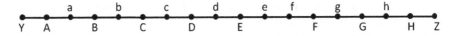

Fig. 3.5 .

A senior citizen after surveying the area decided to stay in a building located along the road YZ, as he found that his friends are already staying in building A, B, C, D, E, F, G and H. The vacant flats are available in all the sixteen buildings on this road. The senior citizen desired to stay in a building at a location such that it minimizes the sum of all its distances from the eight buildings where his friends are staying.

(i)   Can you help the senior citizen to select a building for his stay out of the sixteen buildings mentioned above?

(ii) If his friends staying in building H decide to vacate and shift away from this area, which building will suit the senior citizen so as to minimize the sum of all its distances from the remaining seven buildings where his friends will continue to stay?

35. **Roll Number Dilemma**

When a student in a class was asked his Roll number, he replied as follows:

(i)   If it is a multiple of 4, then my roll number is from 41 to 48.
(ii)   If it is not a multiple of 5, then my roll number is from 51 to 58.
(iii) If it is not a multiple of 8, then my roll number is from 61 to 68.

Find the roll number of the student.

36. **Relative Motion on Circular Track**

Two vehicles X and Y start at 7 AM from location 'A' but travel in opposite directions along a circular track of length 100 km as shown in Fig. 3.6. The speed of X is 1.5 times the speed of Y at the starting point 'A'.

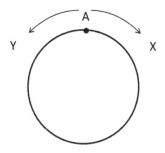

Fig. 3.6 .

Obviously, two vehicles will meet on the circular track at some point. At every meeting point, the two vehicles reverse their direction of travel and interchange their speeds i.e., the speed of vehicle X becomes the speed of vehicle Y and vice-versa.

Find the distance travelled by vehicle X and vehicle Y from starting point 'A' till they both meet again at 'A'.

37. **Race Competition**

Twelve teams participated in a 400 m race competition. It was decided to select the champion based on two rounds of race as follows:

In the first round, the top three teams were selected, where team A finished 10 m ahead of team B and team B finished 15 m ahead of team C.

In the second round, the competition was held among these three teams where the starting point of team A was kept 10 m behind team B and starting point of team B was kept 15 m behind team C as shown in Fig. 3.7. The starting point of team C remains unchanged at starting point of the 400 m line.

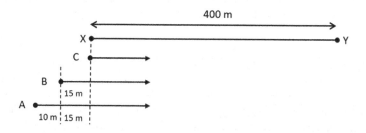

Fig. 3.7 .

Who wins the Race?

Assume that all the three teams run at the same constant speed in both rounds.

38. A Gem from Bhaskaracharya II's Work

Bhaskara II was a great mathematician and astronomer of ancient India in the 12th century AD. The puzzle given below is based on a problem from his book Bijaganita:

"Find two numbers, the sum of whose squares is a cube and the sum of whose cubes is a square".

39. Five Children and 517 Gold Coins

During the excavation of a building foundation, a person unearthed 517 gold coins. He decided to distribute a different number of gold coins among his five children. The number of gold coins each child gets is the integer multiple of gold coins of any other child with a lesser number of gold coins. Also, the least number of gold coins received by any child is more than 1. Find the number of gold coins received by each child.

40. Sum = Product

A mathematician goes to a superstore in a market and buys three items. The store clerk by mistake multiplied the cost of three items instead of adding and obtained 5.49 dollars. When the mathematician paid the amount, the store clerk realized the mistake and told the mathematician that instead of adding the cost of three items, he has multiplied the same. The mathematician told the store clerk, "Don't bother; the addition of three costs also gives the same amount i.e., 5.49 dollars". What are the possible costs of these three items?

41. Investment and Returns

Two friends, one jeweller and another broker decided to try their luck by investing in real estate. So, they jointly purchased a piece of land measuring 3000 square yards. The jeweller invested nine million rupees and the broker invested six million rupees. They divided the entire piece of land into three equal plots each measuring 1000 square yards. Both of them kept one plot each and sold the third plot at a cost of 12 million rupees. Find the amount received by the jeweller and the broker from the sale of the third plot.

42. The Grazing Cow's

Forty cows eat all the grass on 30 acres of field in 10 days and fifty cows eat all the grass on 42 acres of field in 12 days. Find the number of cows which can eat all the grass on 63 acres of field in 15 days.

Assume that when cows start grazing, all the fields are covered with the same quantity of grass per acre of field. Also, the rate of growth of grass is uniform during the period of grazing in all the fields and cows eat the same quantity of grass every day.

43. **Labour on a Construction Site**

On a Bridge construction site, a contractor had employed 50 labourers on the agreement that he will pay 500 rupees to each man, 400 rupees to each woman and 100 rupees to each child per day. It was also made clear that no woman will be allowed unless accompanied by her husband. It was noted that at least half the men working on the site were accompanied by their wives.

If the contractor paid 15000 rupees on a certain day, find the number of men, women and children working on the site on that day.

44. **The Biggest Number Using Three Identical Digits**

Using three 1's, the biggest number obtained is 111 but if we use three 2's, the biggest number obtained is $2^{22} = 4194304$, without using any operational signs. The biggest number obtained using three 9's is given by 9^{9^9}.

So, the biggest number from three identical digits is obtained by a different arrangement of digits depending upon the digit chosen.

Find the biggest number using three 4's without using any operational signs.

45. **Five Jealous Husbands and Their Wives**

Five jealous husbands and their wives have to cross a river in a boat. The boat available is small and can carry a maximum of three persons at a time. Every husband is so jealous that he does not allow his wife to be in company with another man without his presence at any time either on the boat or on the shores.

Find the minimum number of crossings in which five jealous husbands and their wives can cross the river. Assume that boat is not self-driven and at least one person is required to row the boat.

46. **The Look and Say Sequence**

Find the next two terms of the following sequence.

0, 10, 1110, 3110, 132110, 1113122110, 311311222110, ...

47. **Thirty-Six Inch Scale with Minimum Markings**

You can measure any integral number of inches from 1 to 36 on a 36-inch scale with markings at every inch. In fact, all these markings are not required. For example, in a 6-inch scale, two markings i.e., at 1 and 4 are sufficient to measure all integral lengths up to 6 inches (assuming ends 0 and 6 are known) as shown in Fig. 3.8.

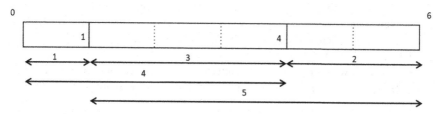

Fig. 3.8 .

Note that measurements shall be taken only once, without lifting and repetition. Find the minimum number of markings required to measure any integral length of inches from 1 to 36 using a 36-inch scale.

48. **Number of Triangles**

The creation of one triangle requires three straight lines if these are neither parallel nor intersecting at a point. Find the maximum number of triangles which can be formed with seven straight lines.

49. **The Ants on a Stick**

Seven ants are walking along the edge of a 1 m stick. Each ant is walking either to the left or to the right at a uniform speed of 1 cm/second. Whenever any ant encounters another ant, both turn around and walk in the reverse direction and whenever an ant reaches the end of the stick, it falls off.

Out of seven ants, four ants namely A, B, D and F walking to the right starts from 5 cm, 10 cm, 35 cm and 63 cm respectively from the left end of the stick. Three ants namely C, E and G walking to the left, starts from 25 cm, 50 cm and 80 cm respectively from the left end of the stick as shown in Fig. 3.9.

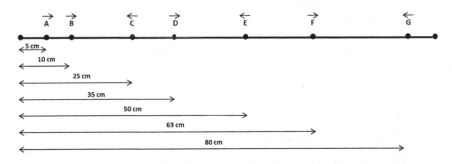

Fig. 3.9 .

Find the following:

   a. Number of encounters that will occur between any two ants.
   b. Maximum time taken by the last ant to fall off.
   c. The last ant to fall off the stick.

50. **Winner of the Year**

Several employees of a company gathered together on 1st January to celebrate the New Year. These employees were divided into two teams say A and B to play a game which will decide the winner of the year.

The rules of the game are:

First team A will select a date in January. Then team B can select a later day either retaining the date and changing the month or retaining the month and changing the date but the day must be later than that selected by team A. Again, team A selects a later day as per the given conditions and this continued till any team is able to select the day as 31 December, who is declared as the winner of the year.

For example, if team A selects 10 January, then team B can select any date from 11 to 31 January or date 10 of any month from February to December. Find the date and month which are selected by a winning team at his turn to ensure its victory.

51. Careless Bank Cashier

One day, I went to a bank to cash my cheque amounting to x rupees and y paisa. The bank cashier by mistake gave me y rupees and x paisa. I did not notice it initially. On the way back, I gave 20 paisa to a beggar. Thereafter, I noticed that I was still having twice the amount of my original cheque. Find the smallest possible amount of the original cheque.

Note: 1 Rupee = 100 paisa.

52. Five Working Days a Week

In a certain organization, all Saturdays, Sundays are holidays and the remaining five days are all working days. If the number of working days is the same in March and April of any year, on what day of the week did April 1st falls in that year.

53. Defective Coin Among 24 Coins

You are given 24 coins including one defective coin. Except for the defective coin, which is lighter than the others, all other coins are of the same weight. Find the minimum weighings required to identify the defective coin on a balance scale without using weights.

54. Two Trains and a Bird

Two trains A and B, 500 km apart start travelling towards each other on a straight track at a uniform velocity of 100 km/h and 150 km/h respectively. At the same time, a bird begins to fly from the face of train A toward train B at a uniform velocity of 200 km/h. After touching the face of train B, the bird reverses direction instantaneously and flies back towards train A with the same velocity of 200 km/h. The bird continues flying back and forth between train A and train B until the two trains collide crushing the bird. Find the total distance travelled by the bird before the crash.

55. Basket and Eggs

The Indian mathematician Brahmagupta during the 7th century AD posed the following problem:

When eggs in a basket are taken out 2, 3, 4, 5 and 6 at a time, there remain 1, 2, 3, 4 and 5 eggs respectively. When they are taken out 7 at a time, none are leftover.

Find the smallest number of eggs that could have been contained in the basket.

56. Ranking Large Numbers

Find the smallest and largest number among the following three numbers,
1000^{1000}, 1001^{999}, 999^{1001}

57. A Unique Polydivisible Number

Find a ten-digit number that uses all the digits 0 to 9 exactly once such that the first
n-digit number is divisible by n for all values of n from 1 to 10 i.e., the first 2-digit
number is divisible by 2, the first 3-digit number is divisible by 3 and so on.

58. Handless Clock Chimer

There is a clock which has a chimer but does not have any hands (i.e., Hour,
Minute or Second). At 1 o'clock, the clock chimes (strikes) once, at 2 o'clock,
it strikes twice and so on. The time gap between any two strikes is constant and
time taken in striking may be taken as negligible.
This clock takes 3 seconds to strike 3 o'clock. How many seconds will it take
to know from the first strike you heard, if it is 10 o'clock?

59. Mango Sharing

Three persons A, B and C fond of mangos (Alphonso variety) jointly ordered
one full box of mangos. During the Covid-19 lockdown period, it took some
time before the sanitized box of mangos was received at the residence of person
A. Soon after taking the delivery, person A opened the box and noted that one
mango was showing signs of rottenness, so he threw that piece and took out his
share i.e., one-third of the remaining pieces which was a whole number.
Sometimes later person B visited person A's residence and threw two of the
rotten mangos and took away one-third of the remaining mangos, again a whole
number. Lastly, person C came to the residence of person A, looked at the
remaining mangos and threw three of the rotten ones and took away one-third
of the remaining mangos, again a whole number. Person A noted that still some
mangos are left, which he kept himself.
Find the minimum number of mangos required to be ordered such that a whole
number of mangos are received by all the persons.

60. Five Sundays in February

The month of February 2020 contained five Saturdays. In the next 100 years,
find the years in which the month of February will contain five Sundays.

61. Employment During Covid-19 Crisis

Many companies decided to reduce the number of employees during the
covid-19 epidemic.
One company decided to reduce the number of its employees considering their
annual performance. They graded their employees as follows:

Grade A	Excellent
Grade B	Very good
Grade C	Good
Grade D	Average
Grade E	Poor

The total strength of the employees in the company is less than 1000.

Exactly one-eighth of the employees got a grade A, one-fifth got a grade B, one-third got a grade C, one-seventh got a grade D and the remaining employees got a grade E.

The company decided to terminate the services of grade E employees.

Find the number of employees who get grade E in their annual performance.

62. **Camel and Banana Transportation**

A banana farmer wanted to transport 2700 bananas to the market, which is 900 km away from his farm. The route passes through the desert, so he decided to use his only camel to transport the bananas.

The camel can carry a maximum of 900 bananas at a time. Find the largest number of bananas which can be transported to the market if

a. Camel needs to eat one banana for every km it travels in loaded condition only.
b. Camel needs to eat one banana for every km it travels whether loaded or empty condition.

The loaded condition means the load of bananas only. Assume that you can load/unload any number of bananas as you want anywhere on the route.

63. **Distance Between Two Colonies**

Five colonies namely A, B, C, D and E are connected by a road network in a city. The distance between colonies A and B is 3 km, between colonies B and C is 7 km, between colonies C and D is 5 km, between colonies D and E is 6 km and between colonies E and A is 21 km.

Find the distance between colonies A and D.

64. **Bachet's Weights**

Find the set of least number of weights required to weigh every whole number of kilograms from 1 to 100 inclusive, if

a. The weights are placed in one scale pan only.
b. The weights can be placed in either scale pan.

65. **Match Stick and Squares**

The 17 match sticks form 6 squares as shown in Fig. 3.10.

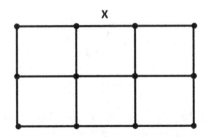

Fig. 3.10 .

a. Can you remove five match sticks so that only 3 squares remain?
b. Remove the top middle match stick marked as X in Fig. 3.10 so that only 16 match sticks remain which form five squares. Can you form exactly four squares by changing the position of three match sticks?

66. Insufficient Pounds

For an important occasion, Amit decided to gift six of his friends with different items. He went to the market to buy six different items with a certain number of pounds in his pocket. On enquiring about the price of items, he found that he was short of money and could only buy any combination of five items out of six at a total cost of 33, 36, 42, 43, 47 or 49 pounds. Find the cost of each of the six items.

67. Mixing Water and Milk

There are two jars, one is a water jar containing 1000 ml of pure water and the other is a milk jar containing 1000 ml of pure milk. Transfer 100 ml of water from the water jar into the milk jar. After stirring, transfer 100 ml of water-milk mixture from the milk jar back into the water jar so that both jars again contain the same volume of 1000 ml as it was at the beginning except for the fact that now both jars contain a mixture of water and milk instead of pure water and pure milk.

Find whether there is more water in milk in the milk jar or more milk in water in the water jar. Assume that the mixing of two liquids does not change the chemistry of either of them.

68. The Three Jug-Decanting Liquid

(i) There are three jugs with capacities of 12 liters, 7 liters and 5 liters. The 12-liter jug is full of milk and the other two jugs are empty. None of the jugs has any markings on them. There is no additional measuring device available.

Find the least number of decanting's (pouring's) required to divide milk in a 12-liter jug into two equal parts of 6 liters each (leaving the smallest jug empty).

(ii) If instead of a 12-liter jug, there is a milkman carrying milk in his 100-liter tank, then find the least number of decanting's (pouring's) in which the milkman can supply 6 liters of milk using 7 liters and 5 liters empty jugs. To avoid wasting milk, the milkman is permitted to pour it back into his 100 liters tank from 5 liters or 7 liters jug if required.

69. Crack the 3-Digit Lock Code

A numeric lock has a three-digit code key. Can you crack the code using the clues given below?
Clues:

A. | 5 | 6 | 3 | One number is correct and placed in the correct position.

B. | 5 | 1 | 8 | One number is correct but placed in the wrong position.

C. | 4 | 9 | 6 | Nothing is correct.

D. | 3 | 0 | 5 | Two numbers are correct but both numbers are placed.
in the wrong position.

E. | 4 | 6 | 0 | One number is correct but placed in the wrong position.

70. Gold and Silver Bricks

During excavation in the field, a farmer unearthed three gold bricks and two silver bricks, which he showed to his three sons. He then selected three bricks at random and put one brick each in three empty lockers namely E, M and Y. He assigned locker E to the eldest son, M to the middle son and Y to the youngest son. None of the three was permitted to see the content of their locker. However eldest son was permitted to see the content of lockers M and Y, the middle son was permitted to see the content of locker Y only but the youngest son was not permitted to see the content of any locker.

The farmer told his sons that whoever tells the content of his locker correctly, is permitted to take the locker with him.

Replying to this, the eldest son says, "I do not know the content of my locker". Then the middle son says, "I also do not know the content of my locker". Finally, after listening to these two statements, the youngest son says, "I now know the content of my locker". So, he was permitted to take his locker with him.

Can you find the content of locker Y?

71. Three Men and Their Wives

Three married couples go to the market and each person buys as many apples as he or she pays in rupees. Each man spends 45 rupees more than his wife. The men are named A, B and C. The women are named X, Y and Z.

Man A buys 17 more apples than woman Y. Man B buys 7 more apples than woman Z. Find the name of each man's (A, B and C) wife.

72. Hundred Balloons in Hundred Rupees

Rishik buys 100 balloons for 100 rupees. The balloons include at least one green, one red, one yellow balloon and no other type. If one green balloon cost 5 rupees, one yellow balloon cost one rupee and 20 red balloons cost one rupee, how many balloons of each colour did he buy?

73. Counterfeit Coin

There are 12 coins including one defective coin. All coins are of the same weight except the defective coin which has a different weight. Find the minimum weighings required to detect the defective coin along with its defect i.e., whether lighter or heavier than the other coins, using a balance scale without using weights.

74. Effect of Paper Folding

Consider a sheet of paper with a size of 1000 mm × 1000 mm and a thickness of 0.1 mm. On folding this paper in half, its dimensions become 500 mm × 1000 mm × 0.2 mm. One more folding in half will change its size to 500 mm × 500 mm × 0.4 mm. Repeat the above folding process. How many times, it is possible to fold the paper. To get a feel of this, estimate the size of paper after folding 10 times, 20 times, 30 times and 50 times.

75. Seven Security Posts

On the circular periphery of a city measuring 39 km, security posts are to be constructed at seven places i.e. A, B, C, D, E, F and G as shown in Fig. 3.11, along the circular road on the periphery.

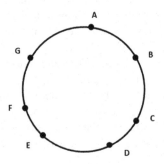

Fig. 3.11 .

The security guards from their posts can walk in either direction to other security posts. The inter distance between security posts is to be fixed in such a way that any integral length from 1 to 38 km can be travelled on the periphery between any two security posts in either direction.

Find the inter distances between security posts.

76. **Rotating Balls**

A mathematician asked his students to look towards a glass chamber in which a number of coloured balls were rotating at a high speed. He informed that the number of black balls lies between 75 and 80% of the total balls rotating in the glass chamber.

Find the smallest possible number of black balls and also the total number of balls rotating in the glass chamber.

77. **Holidays During Covid-19 Crisis**

During the lockdown period of the covid-19 crisis, only limited staff was permitted to attend the office for specified days. A person goes for a 6 km morning walk every day during the holidays. On asking how many km he could afford to go for a morning walk during the days when he attended office also, he replied, "During the lockdown period of 8 to 9 weeks, whenever I attended office, I could go for morning walk only for one-third of my average for the whole period of lockdown".

Find the number of holidays he enjoyed during the lockdown period.

78. **Making the Chain from Cut Pieces**

A jeweller had 40 gold links of old chain in 10 pieces as shown in Fig. 3.12.

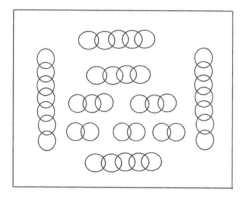

Fig. 3.12 .

He decided to make a single chain by cutting links and soldering. The cost of cutting and soldering a link is 5 dollars.

Find the minimum cost required to be incurred in joining the ten pieces for making

(i) A single endless (close-ended) chain of 40 links.
(ii) A single open-ended long chain of 40 links.

79. Four Operations on Two Numbers

Consider two whole numbers x and y such that x > y.
Perform the following four operations on x and y.

(i)   Add the two numbers.
(ii) Subtract the smaller number from the larger one.
(iii) Multiply the two numbers.
(iv) Divide the larger number by the smaller one.

If the sum of the results of the above four operations is 243, find the two whole numbers.

80. Who'll Become a Decamillionaire (Kaun Banega Crorepati)?

In a game of who'll become a Decamillionaire (Kaun Banega Crorepati), there are two competitors i.e., person A and person B. The organizer stated the rules as follows:

(i)   The game will be started by the person who wins the toss. The initial number i.e., 1 is given to him. Assume that person A wins the toss.
(ii) Multipliers are any integers from 2 to 10.
(iii) The person A is asked to multiply the initial number given to him i.e., 1 by any multiplier from 2 to 10.
(iv) The resultant product is passed on to person B, who is asked to multiply this resultant product by any multiplier from 2 to 10.
(v)   The resultant product is now passed on to person A and the process of multiplying the resultant product by any multiplies from 2 to 10 is repeated.
(vi) Any person who first gets a product equal to or more than 10^7 i.e., one decamillion (i.e., one crore) becomes the winner i.e., decamillionaire (crorepati).

Find the person who can become a decamillionaire (crorepati) i.e., who wins the toss or who loses the toss. Also, find the multipliers which the winner chooses to ensure his win.

81. India's Independence Day

Find the day of the week, when India's Independence Day i.e., 15th August was celebrated in a certain year in which there were exactly four Mondays and four Thursdays in August.

82. Palindromic Mileage

Tarun started his journey at 9:30 AM from Jaipur to Delhi (approximately 300 km apart) by Car. Exactly at 10:18 AM, he noted that the car milometer reading (in km) was 78987, a palindromic number which reads the same from both ends. He was curious to know about the next palindromic number on the car milometer, so he was continuously monitoring the time and milometer reading. To his pleasant surprise, he found the next palindromic number on the car milometer exactly at 11:58 AM.

Find the average speed of the Car.

83. Time Measurement with Hourglass

You have a 4-minute hourglass and a 7-minute hourglass. Find the fastest way to measure 9 minutes using the above two hourglasses.

84. Smallest Product of Two Numbers

Using each of the 10 digits from 0 to 9 exactly once, find the two five-digit numbers such that their product is the smallest possible. Note that 01234 is a four-digit number and not considered a five-digit number. In other words, leading zero is not permitted.

85. Squares and Rectangles on a Chessboard

On a standard 8 × 8 chessboard, find the

(i) Number of squares
(ii) Number of rectangles which are not squares

86. New Year Party and the Poisoned Drink

A five-star hotel organising a New Year party for its guests arranged 120 bottles of drinks consisting of wine/beer/cold drinks etc. and kept them in a secured place in the hotel. An outside miscreant entered the hotel and poisoned one of the bottles of drink. However, the security personnel of the hotel caught the miscreant who informed that he has poisoned only one bottle of drink but does not remember as to which bottle has been poisoned. He also informed that poison is very strong and has a killing effect in 8 to 12 hours time.

Instead of wasting all bottles of drink, the hotel management decided to identify the poisoned bottle in 12 hour's time using Rats as testers. Find the minimum number of Rats required to find out the poisoned bottle of drink. Assume that Rat will die only if fed from the poisoned bottle.

87. Black and White Balls

There are three boxes containing black and white balls. In Box A, there are three times as many white balls as black balls. In Box B, there are six times as many black balls as white balls. In Box C, there are two times as many white balls as black balls.

If the total number of white balls and black balls is 50 each, find the number of white balls and black balls in each of the three boxes.

88. Missing Page Numbers

Some consecutive leaves are torn from a book with page numbers from 1 to 100. If the sum of the remaining page numbers is 5005, find the number of leaves torn and their page numbers. Note that a leaf is a single sheet of paper containing two consecutive page numbers.

89. How Many Byes

In a single-elimination tournament also called knockout, every losing player is eliminated from the subsequent round of play until a single winner is determined. A hundred players participate in a single-elimination tournament of Badminton.

(i)   Find the number of rounds to be played to decide the winner.
(ii) Find the minimum number of byes required to be given.

90. Hourly Consumption of Lemon Juice

Hemant went out on a hot summer day. He kept a 500 ml pure lemon juice in a bottle for consumption during the day.
In the first hour, he drinks 50 ml of pure lemon juice from the 500 ml bottle and refills the bottle with 50 ml of water. In the second hour, he drinks 100 ml of the lemon-water mixture from the 500 ml bottle and refills the bottle with 100 ml of water. Every hour he repeats the same, so in the ninth hour, he drinks 450 ml lemon water mixture from the 500 ml bottle and refills the bottle with 450 ml of water. Finally, in the tenth hour, he drinks 500 ml of the lemon-water mixture and emptied the bottle.
Find the quantity of water consumed by Hemant during the day.

91. Unique Four Digit Square

Rakesh while walking on a road, met with an accident by a cab. At the police station, when enquired about the cab number, Rakesh replied that the cab number was a four-digit perfect square, whose first two digits are the same and the last two digits are also the same. Find the cab number.

92. The Age of Three Children

Dinesh invited his school friend Suresh for a dinner after a long time. During discussions, Suresh asked Dinesh, "How old are your three children". Instead of answering directly, Dinesh replied that the product of their ages is 96 and the sum of their ages is my house number.
Suresh went away to check the house number. After returning, Suresh told Dinesh, "I am still not able to deduce the ages of your three children". Can you give me some more clues?
Dinesh replied, "Your eldest son is older than all my children". He also confirmed to Suresh that now he will certainly be able to find the ages of my children. Suresh replied in the affirmative. Find the house number and the ages of the three children of Dinesh. Assume that all ages are in the whole number of years only.

93. **Pairs of Friendly Rectangles**

A pair of rectangles with sides (a, b) and (c, d) can be considered to be friendly, if the area of each rectangle is equal to the perimeter of the other. Find pairs of friendly rectangles, whose all sides i.e., a, b, c and d are positive even integers.

94. **When RIGHT Becomes LEFT**

Find the smallest integer whose rightmost digit becomes its leftmost digit (without changing any other digit or its position) when multiplied by 8. For example, when an integer 142857 is multiplied by 5, its rightmost digit 7 becomes the leftmost digit as given below:

$142857 \times 5 = 714285$

95. **Seventh Root**

The number 51676101935731 is the seventh power of an integer n. Find the value of n.

96. **Motion on a Straight Road**

Consider a 150 km long straight road AB as shown in Fig. 3.13. Two vehicles X and Y start in opposite directions at the same time from A and B respectively. The speed of vehicle X is 90 km/h and the speed of vehicle Y is 60 km/h.

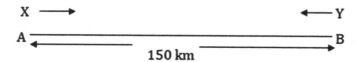

Fig. 3.13 .

Whenever the two vehicles meet, they interchange their speed and reverse their direction of motion. However, if any of the vehicles reach the endpoint A or B, it reverses its direction of motion without any loss of time or change of speed. Find the distance travelled by vehicle X and vehicle Y till they both meet at point B.

97. **Unlucky Thirteen**

Find all possible positive integers which are equal to thirteen times the sum of their digits.

98. **Doubly True Alphametic**

Solve the following alphametic equation and determine the digits represented by each letter. Each letter represents a unique digit from 0 to 9 and leading zeroes are not permitted.

$$FORTY + TEN + TEN = SIXTY$$

99. Tickets for the New Railway Stations

On a certain branch line, there are N railway stations. Every station sells tickets to every other station. Since the distance between certain stations was large, it was decided to construct some new stations. On addition of M new stations, 62 different kinds of new tickets had to be printed.
Find the original number of railway stations N and the number of newly added stations M.

100. Egg Dropping

You are given two identical eggs and access to an 80-story building. On dropping an egg from any floor of the building, the egg will either break or survive. If the egg survives, it can be reused.
Determine the highest floor from which an egg can be dropped without breaking and find the minimum number of drops (trials) in which it can be guaranteed. It can be assumed that if an egg breaks when dropped from the nth floor, then it will also break if dropped from (n + 1)th floor or higher floors. Similarly, if an egg survives if dropped from the nth floor, then it will also survive if dropped from (n − 1)th or lower floors.

101. How Many Squares

Figure 3.14 shows 25 dots in a square pattern of size 4 units × 4 units.

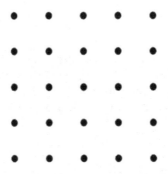

Fig. 3.14 .

How many squares can you locate with dots as their corners?

102. Identify Gold Coin Box

There are three boxes, namely a green box, a white box and a black box. Out of these three boxes, one box contains gold coins and the other two boxes are empty. However, it is not known which box contains gold coins. A statement is written on top of each box as shown in Fig. 3.15.

Green Box	**White Box**	**Black Box**
The gold coins are not in white box.	The gold coins are not in this box.	The gold coins are in this box.

Fig. 3.15 .

Out of the three statements written on the top of the boxes, at least one statement is true and at least one statement is false.

Can you identify the box, which contains the gold coins?

103. Taxi Charges from Jaipur to Delhi

I hired a taxi at the rate of 10 rupees per km and left at 8 AM from Jaipur to Delhi. I told the taxi driver to stop for tea at Neemrana, which is on the way from Jaipur to Delhi. After travelling for two hours, I asked the taxi driver, "How far we had already travelled". The taxi driver replied, "Just twice the distance from here to Neemrana". After travelling for 100 km from here, stopping and leaving Neemrana on the way, I again asked the taxi driver, "How far Delhi is remaining now". He replied, "Just twice the distance from Neemrana to here".

I reached Delhi before Lunch. Find the amount to be paid for the taxi charges.

104. Cheryl's Birthday

Albert and Bernard just became friends with Cheryl and they want to know when her birthday is? Cheryl gives them a list of ten possible dates for her birthday.

May 15	May 16	May 19
June 17	June 18	
July 14	July 16	
August 14	August 15	August 17

Cheryl announced that she will separately tell Albert, the month of her birthday and tell Bernard the day of her birthday. She did it accordingly. Albert and Bernard now have the following conversation.

Albert:	"I don't know when Cheryl's birthday is but I know that Bernard does not know too"
Bernard:	"At first I did not know when Cheryl's birthday is, but I know now"
Albert:	"Now I also know when Cheryl's birthday is"

When is Cheryl's birthday?

105. Twelve Dots and Connecting Lines

Twelve dots are arranged in a rectangular form divided into six squares as shown in Fig. 3.16. Find the minimum number of straight lines required for connecting all twelve dots, passing through each dot only once, without lifting your pen from the paper but ending up at a point where you started.

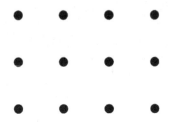

Fig. 3.16 .

106. Equalize Liquid Concentration

There are two jars, one milk jar containing 1000 ml of milk and the other water jar containing 1000 ml of water. Transfer 100 ml of milk from the milk jar into the water jar. After stirring, transfer 100 ml of water-milk mixture from the water jar back into the milk jar so that both jars now contain the same volume of 1000 ml as it was at the beginning. After stirring, continue the process of transferring 100 ml of the liquid mixture from one jar to the other. Find the number of transfers required to equalize the percentage of milk in each jar.

Assume that the mixing of two liquids does not change the chemistry of either of the liquids.

107. Two Boats and Width of a River

Two boats on opposite shores i.e. A and B of a river, start travelling at the same time across the river at right angles to the shore. Both boats travel at a constant speed, though one is faster than the other. These boats meet at 400 m from shore A. On reaching the opposite shore, they turn around and on their return journey, they meet at 200 m from shore B. How wide is the river?

108. First Day of a New Century

On 1st January 2001, I attended office being Monday. People around the world have enjoyed New Year's Day on Sunday many times but the world is waiting for the Sunday which falls on the first day of a New Century.

When can the people enjoy New Century Day on Sunday based on the present Gregorian calendar?

109. Largest Product from Given Sum

Find the largest possible product of positive integers whose sum is 64.

110. Ten Digit Self-descriptive Number

Find a 10-digit number such that the first digit (leftmost) indicates the total number of zeros in the entire number, the second digit indicates the total number of 1's, the third digit indicates the total number of 2's and so on. The last digit i.e., the tenth digit indicates the total number of 9's in the entire number.

An example of such a five-digit number is 21200, where the first digit 2 represents the number of zeros, the second digit 1 represents the number of 1's, the third digit 2 represents the number of 2's, the fourth digit 0 represents the number of 3's and fifth digit 0 represent the number of 4's in the entire number i.e., 21200.

111. Upward Journey of a Ball

A ball is dropped from a height of 200 m from the ground. This ball being elastic hits the ground and rebounds 20% of its previous height on each rebound. Find the total distance travelled by the ball in its upward journey only, before it comes to rest.

112. Mathematician on the Escalator

A mathematician and his wife decided to find the total number of steps of a moving escalator. So, they decided to walk up a moving escalator. The mathematician starts walking twice as fast as his wife on the moving escalator. The mathematician reached the top after climbing 28 steps whereas his wife reached the top after climbing 21 steps. Find the total number of steps on the escalator. Assume that the escalator is moving at a constant speed.

113. Incorrectly Labeled Boxes

There are three boxes, one containing only gold coins labelled as G, another containing only silver coins labelled as S and the third containing mix of gold and silver coins labelled as GS. Somebody has changed the labels in such a way that every box is now incorrectly labelled. In how many trials, you can find the correct labels of boxes. Taking out one coin from any of the boxes will be considered as one trial.

114. How Long Professor Walked

The professor used to finish his office work so as to reach pickup point A at 18.00 hours every day. His son who drives his car from home in time reaches pick up point A exactly at 18.00 hours, picks up his father and drives back home.

One day professor finished his work early and reached pickup point A earlier. He decided to walk home on the same route followed by his son. He meets his son on the way. At meeting point B, the son picks him up and reaches home 30 minutes earlier than usual.

(i) Find the time when the professor meets his son on the way i.e., at meeting point B.

(ii) If the professor reaches pickup point A at 16:30 hours, how long does he walk from pickup point A to meeting point B.

Assume that the son drives the car with constant speed on both sides. The time lost in picking up and turning back may be neglected.

115. Ratio of a Number to its Digit Sum

Find the largest possible value of the ratio of a four-digit number to the sum of its digits.

116. Financial Assistance to Villagers

There are 100 families in a village. During the Covid-19 crisis, it was decided to distribute 1000 rupees to one member of each family in the village, keeping a social distance of 6 feet. Locker with money amounting to 100000 rupees was kept at one place, say A. One member from each family was called and asked to stand up in a row keeping a distance of 6 feet, starting from 6 feet from A. The collector was asked to distribute the amount with the condition that he starts from the place of the locker containing money and finally returns to this place.

Find the distance covered by the collector in distributing money to villagers if:

(i) He can withdraw 1000 rupees only at a time.

(ii) He is permitted to withdraw 1000 rupees at a time till he distributes the money to the first 50% of the persons. After that, he was permitted to withdraw 2000 rupees at a time till he completes distribution and returns back.

117. Gold Coins and Culprit Jeweller

A person purchased 100 gold coins, ten each from the 10 jewellers namely J1, J2, J3, J4, J5, J6, J7, J8, J9 and J10. Every gold coin was supposed to be of a uniform weight of 100 grams each. However, it was later found that one jeweller supplied lighter gold coins each weighing 99 grams, which could not be distinguished visibly from actual gold coins. How many weighing's are necessary to identify the jeweller who supplied lighter gold coins using a digital scale.

118. Pricing of a Fruit Set

Anurag and Rajiv went to the fruit market. Anurag purchased 7 apples, one mango and 19 bananas at a total cost of 230 rupees. Rajiv purchased 6 apples, one mango and 16 bananas at a total cost of 200 rupees.

Find the combined cost of one apple, one mango and one banana.

119. Perfect Powerful Number

Find the smallest positive integer n, such that one half of n is a perfect square, one-third of n is a perfect cube, one-fifth of n is a perfect fifth power and one-seventh of n is a perfect seventh power.

120. Minimum Number of Houses

In a small colony, houses along with a security post are serially numbered from 1 to n in a row. The security post numbered as m exists in between the houses such that the sum of the house numbers on one side of the security post i.e., from 1 to m − 1 is equal to the sum of the house numbers on the other side of the security post i.e., from m + 1 to n. Find the minimum number of houses in the colony.

121. Locker and Coins

A person keeps an even number of gold coins in one locker and an odd number of silver coins in another locker. How to identify which locker contains gold coins.

122. Sum of Digits of Numbers

Find the sum of the digits appearing in the integers 1, 2, 3, 4 ... up to $10^n − 1$.

123. Number of Diagonals in a Polygon

A convex polygon has 65 diagonals. How many sides does the polygon have?

124. Peculiar Pages in a Book

A book has N pages numbered from 1 to N. The total number of digits in the page numbers is 792. How many page numbers of the book contain the digit 1?

125. Coin Distribution

Ten thousand coins are distributed among six houses numbering H1, H2, H3, H4, H5 and H6 in the following manner:
First coin is given to H1, 2nd coin to H2, 3rd coin to H3, 4th coin to H4, 5th coin to H5, 6th coin to H6 and then in backward direction i.e., 7th coin to H5, 8th coin to H4, 9th coin to H3, 10th coin to H2, 11th coin to H1 and then again in forward direction i.e., 12th coin to H2, 13th coin to H3 and so on, till all the coins are distributed in this manner. Find the house which gets the last coin i.e., 10000th coin and the number of coins received by each house.

Chapter 4
Solutions of Creative Puzzles

1. The Economist and a Mathematician

Let's analyse the first proposal of the economist and mathematician.
There are 31 days in January, so the Mathematician is to pay

$$= 1 + 2 + 4 + 8 + \cdots + 2^{31-1} \text{ cents}$$
$$= 2^{31} \text{ cents}$$
$$= 2147483648 \text{ cents}$$
$$\approx 21.475 \text{ million dollars}$$

The Economist offers 0.6 million dollars per day, so the total offer by the economist is

$$= 0.6 \times 31 \text{ million dollars}$$
$$= 18.6 \text{ million dollars}$$

Since 18.6 million dollars is less than 21.475 million dollars, so the Mathematician rejected the offer. Instead, he asked for 0.7 million dollars per day which totals to $0.7 \times 31 = 21.7$ million dollars, which is higher than 21.475 million dollars, however, the offer was not accepted by the Economist. The second month is February and the number of days in February can be 28 in a non-leap year and 29 days in a leap year. Since the discussion between the Economist and the Mathematician is on the first day of the Century, it is apparent that the year is a non-leap year, so February would have 28 days. For 28 days, the Mathematician is to pay

$$= 1 + 2 + 4 + 8 + \cdots + 2^{28-1} \text{ cents}$$
$$= 268435456 \text{ cents}$$
$$\approx 2.684 \text{ million dollars}$$

In return, the Mathematician asked for 0.1 million dollars per day which is equal to $0.1 \times 28 = 2.8$ million dollars for the whole month of February and is more than 2.684 million dollars but agreed by the Economist.
So, the Mathematician gained.

Comments: The apparent reason for losing by Economist is that he might have made a mistake in assuming February of 29 days. For 29 days, the Economist needs to pay only 2.9 million dollars in lieu of payment by the Mathematician which for 29 days will be

$$= 1 + 2 + 4 + 8 + \cdots + 2^{29-1} \text{ cents}$$
$$= 536870912 \text{ cents}$$
$$\approx 5.3687 \text{ million dollars}$$

A Century contains 100 years. 20th Century ended on 31st December 2000 and the 21st Century started on 1st January 2001. So, every New Century starts with the year ending in 01 which will always be a non-leap year.

2. **Mobile Phones and Pen Drives**

This is an interesting puzzle as at first look, it appears that data is insufficient to solve the puzzle.
Let the number of mobile phones be x.
Total sale proceeds $= x \times x = x^2$ dollars
Let the number of 64 GB pen drives be y, the total cost of 64 GB pen drives $= 10 \times y$ dollars. Let the cost of a 32 GB pen drive be z dollars, so

$$x^2 = 10y + z$$

Since the total number of pen drives including one 32 GB pen drive is even, the number of 64 GB pen drives i.e., y is odd.
The cost of a 32 GB pen drive is less than 10 dollars so, $z < 10$.
We have,
$x^2 = 10y + z$, where y is odd and z is a single-digit number.
It is a well-known fact that ten's digit of a perfect square is odd, if and only if, the unit's digit is 6. Refer to Table 4.1.

Table 4.1 .

Units digit	Tens digit
0	0
1	Even
4	Even
5	2
6	Odd
9	Even

Note For a perfect square, the unit's digit can only be 0, 1, 4, 5, 6 and 9

So, if y is odd, $z = 6$

Hence the cost of a 32 GB pen drive = 6 dollars.

Comments: It may be noted that irrespective of the number and price of 64 GB pen drive, the price of a 32 GB pen drive will remain unchanged at 6 dollars, if the number of 64 GB pen drive is odd and the cost of a single 32 GB pen drive is less than 10.

Since x^2 is even, so z can only be 4 or 6

As y is odd so $z = 6$, if y is even then $z = 4$.

3. Inter-hostel Athletics Competition

Important points to note are:

Points for 1st, 2nd and 3rd places are integers x, y and z.

Total points from all events for all teams = 24 + 10 + 10 = 44

There are at least two events i.e., hammer-throw won by Hostel No. 2 and discus-throw.

Overall winner is Hostel No. 1 with 24 points. Hostel No. 2 and Hostel No. 3 got 10 points each.

If n is the total number of events, then

$$n \times (x + y + z) = 44$$
$$= 1 \times 44 = 2 \times 22 = 4 \times 11 = 11 \times 4 = 22 \times 2 = 44 \times 1$$

So, the number of events n can be 1, 2, 4, 11, 22 and 44 with corresponding values of (x + y + z) as 44, 22, 11, 4, 2 and 1 respectively.

Since $x > y > z > 0$, so min value of $z = 1$, $y = 2$, $x = 3$. So, the minimum value of x + y + z = 6, hence n can be 1, 2, or 4 only. The value of n cannot be 1, as we know that there are minimum 2 events i.e. hammer-throw and discus-throw. So, n = 2 or 4.

If n = 2, x + y + z = 22.

If there are 2 events only, then the winning team (Hostel No. 1) must have won a maximum of 1 event as one event (hammer-throw) is won by Hostel No. 2.

Let's consider the points of Hostel No. 2's team:

They have won at least one event. So even if minimum points for 3rd place i.e., z = 1 is considered then maximum points for x = 10 − 1 = 9. But if the maximum point of 1st place x = 9, then Hostel No. 1 team with a maximum of one 1st place cannot have 24 points. Hence n = 2 is not possible.

Let n = 4 and with total point as 44, x + y + z = 11.

Overall champion i.e., Hostel No. 1 team got 24 points. With 4 events, the minimum point for 1st place is 6. Hostel No. 2 team with at least one 1st place got 10 points. So, with 4 events, the maximum points for one 1st place can be 10 − 1 − 1 − 1 = 7. So, 1st place points can either be 6 or 7.

If the 1st place point is 6 then Hostel No. 1 team (winner) must have got 1st place in all four events which are not true as in one event Hostel No. 2 team got 1st place. So, 1st place point can only be 7 so x = 7,

Since x + y + z = 11, so y + z = 11 − 7 = 4 and z < y, so z = 1, y = 3.

Total points of Hostel No. 2 team are 10 with at least one 1st place, so the remaining points for the 3 events are 10 − 7 = 3. Hence in the remaining three events, Hostel 2 team got 1 point only, which is for 3rd place. Hence Hostel 2 team got 1st place for one event and 3rd place for 3 events.

Hostel No. 1 team's total points = 24, which can only be possible with 1st place in 3 events and second place in one event.

As can be seen from Table 4.2, the Hostel No. 3 team got the remaining places in all events i.e., 2nd place in 3 events and 3rd place in 1 event.

Table 4.2 .

Team	Hammer-throw	Discus-throw	Event (3)	Event (4)	Total points
Hostel 1	3	7	7	7	24
Hostel 2	7	1	1	1	10
Hostel 3	1	3	3	3	10

So, the final positions in the discus-throw event are:

1st place	Hostel No. 1 team
2nd place	Hostel No. 3 team
3rd place	Hostel No. 2 team

4. **Ages and Inequalities**

Let the age of the mathematician, his son and daughter be denoted by M, S and D respectively. Now,

$$M + S + D = 100$$
$$M > 2S$$
$$5S > 6D$$
$$3M < 8D$$

From above relations,

$$S < M/2$$
$$D < 5S/6$$
$$M < 8D/3$$

$$\Rightarrow S < M/2 < 4D/3$$
So,

$$100 = (M + S + D) < (8D/3 + 4D/3 + D)$$
$$100 < 15D/3$$
$$\Rightarrow D > 20$$

We also have,

$$M > 2S > 12D/5$$

So,

$$100 = (M + S + D) > (12D/5 + 6D/5 + D)$$
$$100 > 23\,D/5$$
$$\Rightarrow D < \tfrac{500}{23}$$
$$\Rightarrow D < 21.74$$

So,

$$20 < D < 21.74$$

Since D is an integer, so D = 21
Now,

$$M + S + D = 100$$
$$\Rightarrow M + S = 79$$
$$\Rightarrow 79 = (M + S) > (2S + S)$$
$$> 3S$$
$$\Rightarrow S < \tfrac{79}{3} \text{ i.e. } 26.33$$

Now,

$$5S > 6D$$
$$\Rightarrow S > 6D/5$$
$$\Rightarrow S > 6 \times 21/5$$
$$\Rightarrow S > 25.2$$

So,

$$25.2 < S < 26.33$$

For S to be an integer, S = 26
So,

$$M + 26 + 21 = 100$$
$$\Rightarrow M = 53$$

Hence the age of the mathematician is 53 years, his son's age is 26 years and his daughter's age is 21 years. This satisfies all the conditions mentioned.
The solution can also be obtained by plotting the inequalities and the feasible domain will give the required answer.

5. Consecutive Double-Digit Numbers

Important clues in the puzzle are:
The numbers range from 10 (minimum number) to 99 (maximum number) and are consecutive numbers that are known to person X and person Y i.e.

$$X = Y + 1 \text{ or } Y = X + 1, \quad 10 \le X \le 99 \text{ and } 10 \le Y \le 99$$

Initially, person X and person Y do not know the numbers but based on this information and their conversation, both know the numbers.

There are two solutions to the puzzle, depending on whether we start with the minimum number 10 or we start with the maximum number 99.

(a) Let us start with the minimum number 10 and proceed, so, let $X = 10$.

Since 10 is minimum (double-digit number) and Y is consecutive, so Y can be 11 only.

Hence X knows that Y has 11. This contradicts Person X's statement that "I don't know your number". So, X cannot be 10.

Now, if $X = 11$, Y can be 10 or 12.

If $X = 11$ and $Y = 10$, person Y certainly knows that the person X has 11 (same reasoning as given for the case of $X = 10$), so this contradicts Y's statement that "I don't know your number". So, if $X = 11, Y = 12$ is a possibility.

If $X = 12$, Y can be 11 or 13.

If $Y = 11$, X can be 10 or 12 but X cannot be 10 as already reasoned earlier. So, $X = 12$ which is only one possibility so he knows with certainty that $X = 12$ which contradicts his statement "I don't know your number".

Hence for $X = 12$, Y can be 13 only.

So, the second statement in which X says, "I know your number" is satisfied because if $X = 11$, he knows $Y = 12$ and if $X = 12$, he knows $Y = 13$.

But if $X > 12$, again he cannot know with certainty the value of Y, so he cannot make his second statement that "I know your number". Hence the value of X can only be 11 or 12 and the corresponding value of Y can only be 12 or 13 respectively. Hence one of the numbers is certainly 12.

(b) Now start with the maximum number 99 and proceed, so, let $X = 99$.

Since 99 is the maximum (double-digit number) and Y is consecutive, so Y can be 98 only.

Hence X knows that Y has 98. This contradicts Person X's statement that "I don't know your number". So, X cannot be 99.

Proceeding in the same way as detailed in first solution (a), it is easily concluded that one of the numbers, in this case, is certainly 97.

Comments: There can be variations in this puzzle. The consecutive numbers can start from 0 onwards or 1 onwards i.e., 0, 1, 2, 3, 4, 5, ... or 1, 2, 3, 4, 5, ... or any sequence of consecutive numbers with any specified ranges.

For consecutive numbers, if the initial number is A, then the answer to the puzzle is A + 2.

If a range of consecutive numbers from A to B is given, then there are two solutions to the puzzle and the answers to the puzzle are A + 2 and B − 2.

6. Interchangeable Hands of a Clock

Figures 4.1 and 4.2 show a valid time in both the clocks showing an interchange of hands.

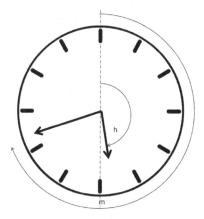

Fig. 4.1 .

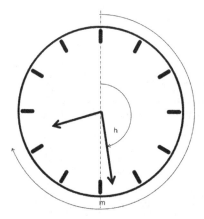

Fig. 4.2 .

Figure 4.1 shows that the hour hand moved h divisions from the 12 o'clock position (let's call the 12 o'clock position as zero position) and the minute hand moved m divisions.

Since the minute hand completes one round in 60 minutes, so let us divide the dial of the clock into 60 divisions. So, the minute hand covers one division in one minute and the hour hand covers 60 divisions in 12 hours. So, hour hand covers 1 division in $\frac{1}{5}$ hours.

So, the hour hand covers h divisions in $\frac{h}{5}$ hours.

Similarly, the minute hand moved m divisions from the zero position.

So, the minute hand covered m divisions in m minutes i.e. $\frac{m}{60}$ hours.

Number of whole hours passed after zero position (i.e., 12 o'clock position)

$$= \frac{h}{5} - \frac{m}{60} = x \qquad (i)$$

Similarly, after hands are interchanged as shown in Fig. 4.2, the number of whole hours passed after the zero position (i.e., the 12 o'clock position)

$$= \frac{m}{5} - \frac{h}{60} = y \qquad (ii)$$

Obviously, $0 \leq x \leq 11$ and $0 \leq y \leq 11$, where x and y are whole numbers. From (i) and (ii),

$$h = \frac{60(12x + y)}{143}$$

and

$$m = \frac{60(12y + x)}{143}$$

With 12 values of x and 12 values of y, we get a total of 144 solutions, but one solution is common to $x = 0, y = 0$ and $x = 11, y = 11$ which shows the zero position (overlapping position). So, in actual, we have 143 solutions covering all required positions of hand which gives valid time for interchanging the hands. For example, let us calculate the positions of hands for $x = 2, y = 3$.

$$h = \frac{60(12 \times 2 + 3)}{143} = 11\frac{47}{143}$$

and

$$m = \frac{60(12 \times 3 + 2)}{143} = 15\frac{135}{143}$$

So, timings are 2 hours $15\frac{135}{143}$ minutes and 3 hours $11\frac{47}{143}$ minutes.

For all combination of values of x and y, where $x \neq y$, we get 66 pairs = 132 solutions and for $x = y$ we get solutions, where minute hand and hour hand overlaps and there are 11 such positions (as $x = y = 0$ and $x = y = 11$ gives the same positions i.e., zero position).

These can be obtained from Eq. (i) by equating $h = m$ where the hour hand and the minute hand coincide. So,

$$\frac{h}{5} - \frac{h}{60} = x \Rightarrow h = \frac{60x}{11}$$

This gives 12 positions for $x = 0$ to 11 where h is at 12 o'clock positions for $x = 0$ or $x = 11$. So, there are 11 distinct positions where the hour hand and the minute hand overlap.

Comments: We have used algebraic method to solve the above puzzle. However, there are other methods to solve the same. Let us discuss the graphical method.

We know that the hour hand completes one round (360°) in 12 hours whereas the minute hand completes one round in 1 hour. The intersecting points of the minute hand line and the hour hand line give positions where the hour hand and minute hand overlap (coincide). There are 11 such intersecting points as shown in Fig. 4.3. Now we interchange the hour hand and the minute hand.

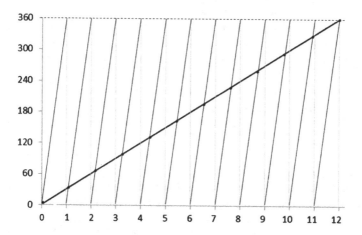

Fig. 4.3 .

Let the hour hand (dark line) become the minute hand line and accordingly scales are also changed as shown in Fig. 4.4. x-axis now shows minutes from 0 to 60 (one complete circle $\approx$ 12 hours). Now draw new lines for the new hour hand as shown by the dashed line. Note that there are 143 points of intersection which gives 143 solutions.

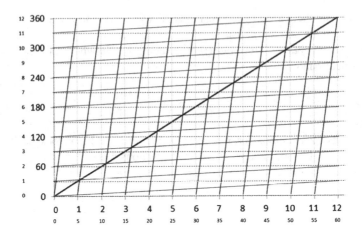

Fig. 4.4 .

7. Largest Product of Two Numbers

For the largest possible product, the first digit of both numbers shall be the largest. Since we have an odd number of digits i.e., 9, so one number will have five digits and the other number will have four digits.

Let the first digit of one number is 9, so the first digit of the other number will be 8. Let the numbers be ABCDE and FGHI. Let A = 8, F = 9 (It will be clear later that placing digit 9 will be more appropriate as the first digit of four-digit number).

The choice for the second digit for both numbers is 7 or 6.

Digit 7 can be at the B position and digit 6 at the G position or vice-versa.

If AB = 87 and FG = 96 then FG − AB = 9.

If AB = 86 and FG = 97 then FG − AB = 11.

Since 9 < 11, the appropriate positions will be given by AB = 87 and FG = 96, because the greatest product is obtained from two numbers if their difference is minimum for the same sum. **This is based on the fact that the area of the square is always greater than the area of the rectangle for the same perimeter**.

Proceeding similarly, the remaining four digits i.e., 5, 4, 3 and 2 can be placed to obtain

ABCD = 8753 and FGHI = 9642.

Now the last digit 1 shall be placed after 3 because (87531 − 9642) < (96421 − 8753).

Hence, we get,

87531 × 9642 = 843973902 and this is the largest possible product.

Comments: Instead of nine digits, if two numbers are to be formed from all 10 digits i.e., 0 to 9 to obtain the largest product, we can proceed in the same way as above and obtain

ABCDE = 87531 and FGHI = 9642

Now the last remaining digit zero can be placed as the last digit of any of the above two numbers to obtain the largest possible product. So, the two solutions are:

87531 × 96420 = 8439739020

875310 × 9642 = 8439739020

Some important observations are:

(i) The first digit must be the largest.

(ii) The value of the digit must decrease from left to right in each number, starting with the larger digit with the number having a smaller leading (first) digit.

(iii) If two numbers are of different sizes, the largest digit must be placed at the start of the smaller number.

If we want to create two numbers from the digits 1, 2, 4, 5, 7, 8 and 9 for the largest possible product, we can easily obtain the two numbers as follows using the observations mentioned above.

First number	Second number
9	8
95	87
952	874
952	8741

The largest product is 952 × 8741 = 8321432.

The largest product of $a < b < c < d < e$ is equal to $dca \times eb$.

8. Three Balls and Four Boxes

Let us summarize the statements and conclusions of the first two persons in a tabular form as shown in Table 4.3, which will clarify the logic behind the conclusions.

Table 4.3 .

Person	Balls drawn	Possible label (incorrect) on the box	Correct content of the box
First	Black-black	BBB BBW	BBW BBB
Second	Black-white	BWW BBW	BBW BWW

The conclusions of correct content of the first two persons as given in Table 4.3 are obvious. For example, if the first person's box is labeled (incorrect) as BBB, then the remaining ball in his box will be white and the correct content of his box (labeled incorrectly as BBB) will be BBW. Since the first two persons correctly identify the content of their boxes, the label (incorrect) of the box assigned to them can only be those given in Table 4.3, which can be easily deduced.

Now let's analyze the statement and conclusions of the third person who draws two white balls and says that he cannot determine the colour of the remaining ball in his assigned box.

With two white balls, the only possibility of box label (incorrect) is BBB or BBW, as in these two cases, he cannot determine the colour of the remaining ball otherwise he can. For example, if the label (incorrect) of the box is WWW, then the colour of the remaining ball will be black only. So, he can determine the correct colour, which violates the given condition that "he cannot determine the colour of the remaining ball in his box".

Similarly, if the label (incorrect) of the box is BWW, then the remaining ball will be white only, so again he can determine the correct colour of the remaining ball in his box.

So now the updated details are shown in Table 4.4.

Table 4.4 .

Person	Ball drawn	Possible label (incorrect) on the box	Correct content of the box
First	Black-black	BBB BBW	BBW BBB
Second	Black-white	BBW BWW	BWW BBW
Third	White-white	BBB BBW	

On observing Table 4.4, it can be seen that none of the first three persons contain the box with the label (incorrect) as WWW, so the fourth person knows this without seeing that his box label (incorrect) is WWW.

Since the fourth person's box label (incorrect) is WWW, so his content of balls cannot be three white balls. From Table 4.4, it can also be seen that the correct content of boxes assigned to the first and second person cannot be three white balls, hence third person's box must contain three white balls.

It can also be observed from Table 4.4 that the second person's box is the only box with the label (incorrect) as BWW so no other person can contain this box. Hence second person's box label (incorrect) is BWW and accordingly the correct content of his box is BBW (i.e., the remaining ball is black). Hence the correct content of the first person's box will be BBB (i.e., the remaining ball will be black) and its label (incorrect) will be BBW. So, the only remaining label (incorrect) of the box is BBB will pertain to the third person. Finally, the fourth person's box will contain the remaining combination of balls i.e., BWW i.e., two white and one black ball.

The final details are shown in Table 4.5.

Table 4.5 .

Person	Balls drawn	Box label (incorrect)	Remaining ball	Correct content of the box
First	Black-black	BBW	Black	BBB
Second	Black-white	BWW	Black	BBW
Third	White-white	BBB	White	WWW
Fourth	–	WWW	–	BWW

9. Worm on a Rubber Rope

At a first glance, it appears that the worm will never reach the other end of the rope because we are tempted to think that the worm and rope are moving independently, which is not so. Since rubber rope stretches uniformly, so, at any time portion of rubber rope BEHIND the worm also stretches in the same proportion as the portion of rubber rope INFRONT of the worm. This means that the worm (even without crawling) is carried forward with every stretching. After one second of crawling, the worm is 1 cm from one end i.e., one percent of rope length and 99 cm i.e., 99% of rope length from the other end as shown in Fig. 4.5.

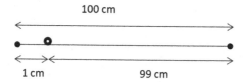

Fig. 4.5 .

Now rubber rope stretches instantly by 1 m, increasing the length of rope to 2 m. Since rubber rope stretches uniformly, the position of the worm will remain at a distance of 1% of the length of rope from one end and 99% from the other end i.e., the worm is now 1% of 2 m = 2 cm from one end and 99% of 2 m = 198 cm from another end of the rope as shown in Fig. 4.6.

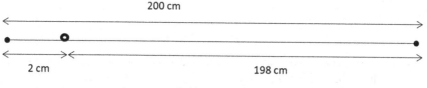

Rubber rope after stretching

Fig. 4.6 .

The key point to note is that though the worm has crawled by 1 cm only but due to the stretching of rubber rope, it has covered 2 cm from starting position.

At the end of 2 seconds, the worm crawls 1 cm more, so the total distance covered by the worm becomes 2 cm + 1 cm = 3 cm from one end (starting position) and 197 cm from far end. That means worm is at $1\% + \frac{1}{2}\% = 1.5\%$ of rope length from starting position and $100\% - 1.5\% = 98.5\%$ from far end. The rubber rope is instantly stretched again after 2 seconds making it 3 m long. As explained above, the worm which is at 1.5% of rope length before stretching will remain at 1.5% of rope length after stretching also. So, the position of the worm will be at a distance of 1.5% of 3 m = 4.5 cm from starting position and $300 - 4.5 = 295.5$ cm from the far end. Similarly, it can be seen that after 3 seconds, the position of the worm will be 4.5 cm + 1 cm = 5.5 cm from starting position and $300 - 5.5 = 294.5$ cm from the far end. In percentage terms, the worm will be at $(1 + \frac{1}{2} + \frac{1}{3})\%$ of rope length from starting position after 3 seconds. So, after n seconds, the distance covered by the worm as a percentage of initial rope length is

$$= \left(1 + \frac{1}{2} + \frac{1}{3} + \frac{1}{4} + \cdots + \frac{1}{n}\right)\%$$

$$= \left(\frac{1}{100} + \frac{1}{200} + \frac{1}{300} + \cdots + \frac{1}{100n}\right)$$

$$= \left(1 + \frac{1}{2} + \frac{1}{3} + \frac{1}{4} + \cdots + \frac{1}{n}\right)/100 \text{ considering } 1 \text{ m} = 100 \text{ cm length of rope.}$$

The series in parenthesis is a well-known sum of the Harmonic series and can be made as large as possible by increasing the value of n. When the value of n is such that the sum of harmonic series inside parenthesis becomes more than 100, the worm will reach the other end of the rope in n seconds. However, the value of n will be very large as can be seen below:

The sum of series in parenthesis can be expressed as

$$= 1 + \frac{1}{2} + \left(\frac{1}{3} + \frac{1}{4}\right) + \left(\frac{1}{5} + \frac{1}{6} + \frac{1}{7} + \frac{1}{8}\right) + \cdots$$

Now $\frac{1}{3} + \frac{1}{4} > \frac{1}{2}$

and $\frac{1}{5} + \frac{1}{6} + \frac{1}{7} + \frac{1}{8} > \frac{1}{2}$ and so on.

So, the sum of the series becomes more than

$$1 + \frac{1}{2} + \frac{1}{2} + \frac{1}{2} + \cdots$$

That can be made as large as possible by increasing the number of terms of the series. It can be seen that sum of the series from 1 to $1/2^n$ exceeds $\frac{n}{2}$. So, the sum of the series from 1 to $1/2^{2n}$ exceeds n.

For n = 100, the sum of series from 1 to $1/2^{200}$ will exceed 100, so the number of seconds after which the worm will reach the far end will be within 2^{200} seconds. More precisely, this comes within e^{100} seconds, where e is the base of the natural logarithm i.e., e = 2.71828....

This number e^{100} is very large and is greater than the age of the universe.

Comments: Let's consider what happens, if instead of 1 m stretching of rubber rope per second, it is stretched by 1 m after one second, 2 m after two seconds, 4 m after three seconds, 8 m after four seconds and so on up to 2^{n-1} m after n seconds. This time the stretching is in Geometric progression.

Proceeding in the same way as explained earlier, the distance travelled by the worm after n seconds.

$$= 1/100 + 1/200 + 1/400 + 1/800 + \cdots + 1/\left(100 \times 2^{n-1}\right)$$
$$= \left(1 + 1/2 + 1/4 + 1/8 + \cdots + 1/2^{n-1}\right)/100$$

The quantity in parenthesis is the sum of a geometric series, the sum of which is less than 2, even if the number of terms i.e., n is increased to infinity, the sum $= 1/(1 - \frac{1}{2}) = 2$.

So, the maximum distance which can be covered by a worm can be 2% of the length of a rubber rope and will never reach the far end of the rubber rope.

10. **Winning Numbers for 101**

   To get a winning total of 101 and more, the opponent (loser) must be given the largest sum S such that $(S + 9) < 101$

$$\Rightarrow S < (101 - 9), \text{so } S = 91$$

   This is the sum to be given to the opponent (Loser) to force him to lose, as he can select numbers from 1 to 9 only to add, which gives the sum from 92 to 100 to the opponent (winner) who can make it to 101 by adding a required number from 1 to 9. So, 91 is the winning sum. Going backward in a similar way, the second largest sum say S1 given to the opponent (loser) at the previous turn is such that $(S1 + 9) < 91$

$$\Rightarrow S1 = 81$$

   So, when the opponent (Loser) gets this sum of 81, he can only add 1 to 9, which gives sums from 82 to 90, which can easily be converted by the opponent (winner) to 91 by adding the required number from 1 to 9 (i.e., if he gets 82, he adds 9, if he gets 90, he adds 1) So 81 is the next winning sum.

   Proceeding backward in a similar way, the remaining winning sums i.e., 71, 61, 51, 41, 31, 21, 11 can be easily calculated. So, if the first player, say A starts with number 1, he can certainly force the opponent player B to lose, by successively making the sum to winning sums as given above. But if player A starts with any other number from 2 to 9, his opponent can manage to win by making the sum to winning sums.

   So, the numbers to be selected by the winner are such that, after starting with number 1, he can make the sum to 11, 21, 31, 41, 51, 61, 71, 81 and 91 at his successive turns.

Comments: This game can be generalized as follows.

Let's call the final number to be attained by the winner (i.e., 101 in above puzzle) as the target Number T. Let the winning sum to be ensured in previous positions is W1, W2, W3 ... and let the numbers permitted to be added are 1 to n.

If the sum W1 given by player A is such that W1 + n is equal to or more than T, then the opponent player i.e., B can add to W1, any number from 1 to n to make his sum to T and win. So, to force the opponent player B to lose, player A must ensure that W1 to be given to player B be such that

$$W1 + n + 1 = T \Rightarrow W1 = T - (n + 1)$$

In this case, opponent player B cannot get the target number T, by adding any number from 1 to n, so he is forced to give his opponent player A, a sum, which player A can make it to T by adding a suitable number from 1 to n. So W1 is the winning sum that guarantees player A to win.

Proceeding backward in the same way, we can calculate winning sums W2, W3 ... as

$$W2 = T - (n + 1) - (n + 1) = W1 - (n + 1)$$
$$W3 = W2 - (n + 1)$$
$$W4 = W3 - (n + 1)$$
$$W_m = W_{m-1} - (n + 1) \text{ till } W_m \geq 1$$

So, if player A ensures his sums to the above winning sums i.e., W1, W2 ..., he can ensure his win irrespective of numbers added by opponent player B.

11. **Keys to Open the Locker Room**

Let the given bank officials be designated as B1, B2, B3 and B4.

It is given that no two bank officials can open the locker room but any group of three bank officials can open the locker room.

Suppose B1 and B2 try to open the locker room, then there must be at least one lock say L1 which they are not able to open. Now if any of the remaining two i.e., B3 or B4 joins B1 and B2, then it is possible to open the locker room. Hence B3 and B4 both must have a key for lock L1. So, lock L1 must have two keys, one each for B3 and B4. Now let B1 and B3 try to open the locker room, then there must be at least one lock say L2, which they are not able to open. If any of the remaining two i.e., B2 or B4 join B1 and B3, then it is possible to open the locker room. Hence B2 and B4 both must have keys for lock L2. So L2 must have two keys, one each for B2 and B4.

Based on the above logic, for every pair, out of 4 bank officials, there is one lock that they are not able to open. The number of pairs of bank officials out of 4 is B1B2, B2B3, B3B4, B4B1, B1B3 and B2B4. So minimum of 6 locks are required i.e., L1, L2, L3, L4, L5 and L6 and the minimum no. of keys required = 2 × 6 = 12.

Let key numbers corresponding to L1, L2, L3, L4, L5 and L6 are 1, 2, 3, 4, 5 and 6. So the distribution of keys among four bank officials can be as

B1:	4, 5, 6
B2:	2, 3, 6
B3:	1, 3, 5
B4:	1, 2, 4

It can be seen that every combination of three bank officials contains keys for all six locks required to open the locker room but none of the combinations of two bank officials contain keys for all the six locks.

Comments: Consider the case of three bank officials B1, B2 and B3 such that no single official can open the locker room, however, any group of two bank officials can open the locker room by pooling their keys. Based on similar logic as explained above, it is easy to conclude that minimum of three locks are required with 6 keys, two keys with each official i.e. B1: 2, 3, B2: 1, 3 and B3: 1, 2.

It is easy to generalize the solution for n bank officials where no group of r officials can open the locker room but every group of r + 1 officials can open the locker room.

The minimum number of locks required = ^nC_r, where ^nC_r corresponds to the number of distinct groups of r people out of n people.

Minimum no. of keys required per person = $^{(n-1)}C_r$

For n = 3, r = 1, min. no. of locks = $^3C_1 = 3$

$$\text{Minimum number of keys per person} = {}^2C_1 = 2$$

For n = 4, r = 2, min. no. of locks = $^4C_2 = 6$

$$\text{Minimum number of keys per person} = {}^3C_2 = 3$$

For n = 6, r = 3, min. no. of locks = $^6C_3 = 20$

$$\text{Minimum number of keys per person} = {}^5C_3 = 10.$$

In this way, minimum number of locks and keys required for any combination of n and r can be calculated.

12. Fastest Three Horses

Since only five horses can be allowed to race at a time, split the 25 horses into groups of five horses each. Let these groups be designated as A, B, C, D, and E. Each group consists of five horses. Run five races i.e., one race for each group of 5 horses. Let the result of the five races is:

Group A:	A1, A2, A3, A4, A5
Group B:	B1, B2, B3, B4, B5
Group C:	C1, C2, C3, C4, C5
Group D:	D1, D2, D3, D4, D5
Group E:	E1, E2, E3, E4, E5

The digits 1 to 5 denote the position of horses in the group, after the first race of their group. So A1 is the fastest and A5 is the slowest in group A. Fastest horses in each group are A1, B1, C1, D1 and E1.

Now 6th race is conducted among the five fastest horses i.e. A1, B1, C1, D1 and E1. Let the result of the 6th race is:

$$A1 > B1 > C1 > D1 > E1$$

So A1 is fastest and E1 is slowest in the 6th race. The following conclusions can be drawn from the results of six races:

(i)   A1 is the fastest of all 25 horses so need not be run again.
(ii) D1 and E1 are the fastest of their groups D and E respectively. Both D1 and E1 are slower than the three horses A1, B1 and C1 so all horses in group D and group E can be eliminated as they are out of competition for the fastest three horses. Similarly, all horses at 4th and 5th position in their respective group are also out of the race.
(iii) Horse B3 can be eliminated as it is slower than A1, B1 and B2.
(iv) Horse C2 and C3 can be eliminated as they are slower than A1, B1 and C1.

The key point is that any horse which is slower than possible top three fastest horses can be eliminated.

So now we are left with the following five horses only i.e. A2, A3, B1, B2 and C1. The final and 7th race can be conducted among these five horses i.e. A2, A3, B1, B2 and C1. The top two horses in this race get the second and third fastest positions.

So, seven races are required to identify the three fastest horses out of 25 horses.

13. **Climbing the Stairs**

Let N denotes the number of steps.

For N = 1, there is only one way to climb up one step i.e. (1).

For N = 2, there are two ways to climb up the second step i.e., either by taking two single steps or one double step i.e. (1, 1), (2).

For N = 3, there are three ways to climb up the third step i.e., either by taking three single steps or first double step followed by a single step or the first single step followed by a double-step i.e. (1, 1, 1), (2, 1) and (1, 2) as shown in Fig. 4.7.

Three single steps
(1,1,1)

First double step
then single step
(2,1)

First single step then
double step (1,2)

Fig. 4.7 .

The same logic can be applied to higher steps.

For N = 4, there are five ways to reach the top of fourth step i.e., (1, 1, 1, 1), (1, 1, 2), (1, 2, 1), (2, 1, 1), (2, 2).

It can be seen that the fourth step can be climbed up either by taking one step at a time i.e., from the third step or two steps at a time i.e., from the second step. So, the number of ways to climbed up the fourth step

$$= \text{Number of ways to climbed up the third step}$$

$$+ \text{Number of ways to climbed up the second step}$$

$$= 3 + 2 = 5$$

If the number of ways to climbed up the Nth step is denoted by F(N) then

$$F(N) = F(N-1) + F(N-2) \tag{i}$$

Because Nth step can be climbed up either by taking one step at a time from $(N-1)$ th step or by taking two steps at a time from $(N-2)$th step. So, as seen earlier

$$F(1) = 1, F(2) = 2, F(3) = 3, F(4) = 5$$

From equation (i), we can get

$$F(5) = F(3) + F(4) = 3 + 5 = 8$$
$$F(6) = F(4) + F(5) = 5 + 8 = 13$$
$$F(7) = F(5) + F(6) = 8 + 13 = 21$$
$$F(8) = F(6) + F(7) = 13 + 21 = 34$$
$$F(9) = F(7) + F(8) = 21 + 34 = 55$$
$$F(10) = F(8) + F(9) = 34 + 55 = 89$$

So, there are 89 distinct ways to climb the staircase with 10 steps.

Comments: It can be seen that number of distinct ways to climb the N step forms the series 1, 2, 3, 5, 8, 13, 21, 34, 55, 89 (for N = 1 to 10). This series can be extended to any term and is the famous Fibonacci sequence. So, the number of distinct ways to climb N steps is given by Nth Fibonacci number F(N). Starting with F(0) = 1 and F(1) = 1, further terms can be obtained by Eq. (i). Another variation of the puzzle can be obtained, if instead of two steps, at most three steps can be permitted at a time i.e., one can climb 1, 2 or 3 steps at a time, then number of distinct ways to climb n steps can be calculated as earlier i.e., for

$n = 1, F(1) = 1$

$n = 2, F(2) = 2$

$n = 3, F(3) = 4[(1, 1, 1), (1, 2), (2, 1), (3)]$

$n = 4, F(4) = 7[(1, 1, 1, 1), (1, 2, 1), (1, 1, 2), (2, 1, 1), (3, 1), (1, 3), (2, 2)]$

So, if at most 3 steps are permitted, then nth step can be climbed up in F(n) distinct ways where,

$$F(n) = F(n-1) + F(n-2) + F(n-3)$$
$$F(5) = F(4) + F(3) + F(2) = 7 + 4 + 2 = 13$$

This can be generalized to if at most m steps are permitted at a time and in that case

$$F(n) = F(n-1) + F(n-2) + F(n-3) + \cdots + F(n-m)$$

14. Shopping During Partial Lockdown

Let n donates the original amount of money (in rupees).
Amount spent at groceries shop

$$= \frac{n}{2} + \frac{1}{2} = \frac{n+1}{2}$$

Amount spent at sweets shop

$$= \frac{1}{2}(\text{remaining amount}) + \frac{1}{2}$$
$$= \frac{1}{2}\left(n - \frac{n+1}{2}\right) + \frac{1}{2}$$
$$= \frac{n+1}{2^2}$$

Amount spent at stationery shop

$$= \frac{1}{2}\left(n - \frac{n+1}{2} - \frac{n+1}{2^2}\right) + \frac{1}{2}$$
$$= \frac{n+1}{2^3}$$

Similarly, we can find that amount spent at fruits and vegetable shop is $\frac{n+1}{2^4}$ and $\frac{n+1}{2^5}$ respectively.

$$\text{Amount donated at temple} = \frac{1}{2}(\text{remaining amount}) + \frac{1}{2}$$
$$= \frac{n+1}{2^6}$$

The remaining amount = 100 rupees. So

$$n = \frac{n+1}{2} + \frac{n+1}{2^2} + \frac{n+1}{2^3} + \cdots + \frac{n+1}{2^6} + 100$$
$$n = (n+1) \times \left(\frac{1}{2} + \frac{1}{2^2} + \frac{1}{2^3} + \cdots + \frac{1}{2^6}\right) + 100$$

We can easily find the sum of the geometric progression $(\frac{1}{2} + \frac{1}{2^2} + \frac{1}{2^3} + \cdots + \frac{1}{2^6})$ appearing in above equation. So, $n = (n+1)(1 - \frac{1}{2^6}) + 100 \Rightarrow n = 6463$.

So, 6463 rupees is the initial amount of money, I was having when I went to the market.

Amount spent at groceries shop $= \frac{6463}{2} + \frac{1}{2} = 3232$ rupees

Amount spent at sweet shop $= \frac{6463-3232}{2} + \frac{1}{2} = 1616$ rupees

Similarly, amount spent at the stationery shop, fruit shop and the vegetable shop are 808, 404 and 202 rupees respectively.

$$\text{Amount donated at temple} = \left[6463 - (3232 + 1616 + 808 + 404 + 202)\right]\frac{1}{2} + \frac{1}{2}$$
$$= \frac{101}{2} + \frac{1}{2} = 101 \text{ rupees}$$

$$\text{Remaining amount} = 6463 - (3232 + 1616 + 808 + 404 + 202 + 101)$$
$$= 100 \text{ rupees by which milk was purchased while returning home.}$$

Comments: At a first glance this puzzle appears to be cumbersome but when a pattern of a geometric progression is seen, the solution appears easy. The concept of the sum of geometric progression can be applied in between also while computing the remaining amount. For example, at the fruit shop, the remaining amount is

$$= n - \left(\frac{n+1}{2} + \frac{n+1}{2^2} + \frac{n+1}{2^3}\right) = n - (n+1)\left(\frac{1}{2} + \frac{1}{2^2} + \frac{1}{2^3}\right)$$
$$= n - (n+1)\left(1 - \frac{1}{2^3}\right) = \frac{n-7}{8}$$

So, the amount spent at the fruit shop

$$= \left(\frac{n-7}{8}\right)\frac{1}{2} + \frac{1}{2} = \frac{n+1}{2^4}$$

If there are spending at many places in a similar fashion, then spending at xth shop $= \frac{n+1}{2^x}$.

15. Shaking Hands

At the dinner party, a total of 12 people are there. Since no one shook their hand or the hand of their spouse, a single person can shake hands with a maximum of 10 persons. Since I received eleven different answers so this implies that one person did not shake hands with any other person. Therefore, eleven different answers by 11 persons where each shook a different number of hands must be 0, 1, 2, 3, 4, 5, 6, 7, 8, 9 and 10.

The person, who had 10 handshakes, must have shaken hands with everyone except his spouse. Let's call this person P_{10}. So, everyone except the spouse of P_{10} had at least one handshake. Hence out of 11 different answers, the reply given as zero must be of the spouse of P_{10}. So, the person who had 10 handshakes, his spouse had zero handshakes. The person who replied 9 handshakes (say P_9) must have shaken hands with everyone except his spouse and spouse of P_{10} (with zero handshakes). So, everyone except the spouse of P_9 and the spouse of P_{10} had handshakes with at least two persons. Since the spouse of P_{10} had zero handshakes, all others except the spouse of P_9 had two or more handshakes, so the spouse of P_9 must have replied one handshake only. We have seen that

P_{10} had 10 handshakes, spouse of P_{10} had zero.

P_9 had 9 handshakes, spouse of P_9 had one.

Proceeding in this way, it can be deduced that P_8 had 8 handshakes, his spouse had two and so on.

So, handshakes by five invited couples are (10, 0), (9, 1), (8, 2), (7, 3) and (6, 4). Now only one reply as 5 handshakes is left which belongs to the left-out person i.e., my wife. So, my wife shook hands with 5 people.

Comments: The puzzle can be generalized as follows:
Let one couple invites N other couples for a party. All other conditions remain the same as given in the puzzle, let's find out how many hands did my wife shake?
Total number of couples = N + 1
Total number of persons present = 2 (N + 1)
The number of different answers received by me (host)
= 2 (N + 1) − 1 = 2N + 1.

Proceeding similarly, the person who replies 2N handshakes must have shaken hands with everybody else except their own spouse, so everybody else had at least one handshake except his spouse who must have replied zero handshakes. Similarly, the pairs of handshakes of persons with their spouses will be (2N, 0), (2N − 1, 1), (2N − 2, 2), … [2N − (N − 1), N − 1]. The last pair is (N + 1, N − 1).
So, all persons who answered from 0 to N − 1 and N + 1 to 2N are covered. The leftover person i.e., the spouse of the host must have replied the left-out number i.e. N. So the spouse of the host shook hands with N persons.

16. Equal Distribution by Doubling

This is an interesting puzzle and a general solution exists for n persons but let us first consider the case of two persons A and B with an initial number of the silver coins as a and b respectively where a > b.
Person A doubles the number of coins of person B which he already held. So, the current distribution will be (a − b) silver coins with Person A and (b + b) silver coins with person B, as the content of person A is reduced by b and the content of person B is increased by b.
Now person B doubles the number of coins of person A, which he currently held. So finally
A has (a − b) + (a − b) = 2a − 2b and B has 2b − (a − b) = 3b − a
Now final content of person A and B is equal, so 2a − 2b = 3b − a
$\Rightarrow 5b = 3a \Rightarrow \frac{a}{b} = \frac{5}{3}$
Hence the smallest value of a = 5 and b = 3, which is the initial content.
Final content of person A = 2a − 2b = 2 × 5 − 2 × 3 = 4
Final content of person B = 3b − a = 3 × 3 − 5 = 4
We can proceed in this way, but it becomes cumbersome for increased number of persons.
The best strategy in such cases is to start backward from the end.
Let the number of silver coins each of the five persons held finally = x.
Before this, person E doubles the number of silver coins of all other persons i.e. A, B, C and D, which resulted in an equal number of silver coins finally with everybody. So before doubling silver coins by E, the number of silver coins held by A, B, C and D must be x/2 and silver coins held by E must be x + 4 (x/2).
Let us put it in a tabular form as shown in Table 4.6 and compute the silver coins held by each person at each stage starting from the end and going back to the initial position as given below.
It can be seen that if the final distribution is equal i.e. x with each of the persons A, B, C, D, and E, then the initial distribution of A, B, C, D, and E is in the ratio of 81:41:21:11:6.
The smallest number of coins with all the persons can be obtained by choosing the smallest value of x i.e., 32 to make all values as an integer number of coins.
So, for n = 5, the smallest number of silver coins initially held by A, B, C, D, and E are 81, 41, 21, 11 and 6 respectively. The final number of silver coins held by each of the five persons will be 32 i.e., 2^5.

Table 4.6 .

S. No.	Stage	Number of silver coins held by a person					Remarks
		A	B	C	D	E	
1	Final	x	x	x	x	x	All have equal no.
2	One stage before final	$\frac{x}{2}$	$\frac{x}{2}$	$\frac{x}{2}$	$\frac{x}{2}$	$x + \frac{4x}{2} = 3x$	Person E doubles the coins of all others
3	Two stages before final	$\frac{x}{4}$	$\frac{x}{4}$	$\frac{x}{4}$	$\frac{x}{2} + \frac{3x}{4} + \frac{3x}{2} = 11x/4$	$\frac{3x}{2}$	Person D double the coins of all others
4	Three stages before final	$\frac{x}{8}$	$\frac{x}{8}$	$x/4 + x/8 + 11x/8 + 3x/4 = 21x/8$	$\frac{11x}{8}$	$\frac{3x}{4}$	Person C doubles the coins of all others
5	Four stages before final	$\frac{x}{16}$	$\frac{x}{8} + \frac{x}{16} + \frac{21x}{16} + \frac{11x}{16} + \frac{3x}{8} = 41x/16$	$\frac{21x}{16}$	$\frac{11x}{16}$	$\frac{3x}{8}$	Person B doubles the coins of all others
6	Initial Stage	$\frac{x}{16} + \frac{41x}{32} + \frac{21x}{32} + \frac{11x}{32} + \frac{3x}{16} = 81x/32$	$\frac{41x}{32}$	$\frac{21x}{32}$	$\frac{11x}{32}$	$\frac{3x}{16}$	Person A doubles the coins of all others

Comments: It is interesting to note that for n persons, the minimum initial content will be as follows:

The person with the smallest number of coins initially will have n + 1 coins. Second person will have 2 × (first person's content) − 1 = 2 (n + 1) − 1

$$\text{Third persons content} = 2 \times (\text{second persons content}) - 1$$
$$= 2 \times [2(n+1) - 1] - 1 \text{ and so on.}$$

If X(n, 1) denotes the smallest number of coins of the first person initially, we have X(n, 1) = n + 1, Then the content of the kth person

$$X(n,k) = 2 \times X(n, k - 1) - 1 \text{ where } 2 \leq k \leq n$$

and final content of each person = 2^n, which is the sum of the initial content of all persons divided by the number of persons.

Table 4.7 gives a solution to the puzzle for the number of persons from 2 to 10.

Table 4.7 .

No. of persons (n)	The initial number of silver coins with n persons
2	3, 5
3	4, 7, 13
4	5, 9, 17, 33
5	6, 11, 21, 41, 81
6	7, 13, 25, 49, 97, 193
7	8, 15, 29, 57, 113, 225, 449
8	9, 17, 33, 65, 129, 257, 513, 1025
9	10, 19, 37, 73, 145, 289, 577, 1153, 2305
10	11, 21, 41, 81, 161, 321, 641, 1281, 2561, 5121

The final content will be 2^n with each of the n persons. The table can be easily extended to any number of persons.

17. Identify the Apple Juice Bottle

Let the quantity of orange juice given to neighbour = x, so the quantity of orange juice given to relative = 2x.

Total quantity of orange juice = x + 2 x = 3x.

So, the total quantity of orange juice must be divisible by 3.

The sum of total quantity of all bottles

$$= 10 + 12 + 16 + 17 + 21 + 22 + 33 = 131$$

Since 131 gives a remainder of 2 when divided by 3, so bottle containing the apple juice will be the one whose quantity gives a remainder of 2 when divided by 3. The only possibility out of seven bottles satisfying this condition is the bottle containing 17 liters, as 17 divided by 3 leaves a remainder of 2. Hence 17 liters bottle contains apple juice.

Comments: Just to identify the apple juice bottle, the number of juice bottles given to the neighbour is not required. However, if given, it is possible to identify the orange juice bottles given to the neighbour and relative.
Quantity of orange juice = 131 − 17 = 114 liters.
The quantity given to the neighbour = 114/3 = 38 liters, so three bottles containing 10, 12 and 16 liters are given to the neighbour.
The quantity given to the relative is the remaining three bottles containing 21, 22 and 33 bottles.
It can be seen that 21 + 22 + 33 = 76 = 2 × 38 liters.
If he has given only two bottles to the neighbour instead of three then bottles containing 16 liters and 22 liters (i.e., 16 + 22 = 38 liters) are given to the neighbour. The remaining bottles i.e., 10, 12, 21 and 33 liters are given to the relative.
It can be seen that 10 + 12 + 21 + 33 = 76 = 2 × 38 liters.

18. Lockers Finally Remain Opened

To simplify, let us first consider 10 lockers only. The status of open/closed lockers at every step is tabulated in Table 4.8, where the closed position is shown by C and the open position by O.

Table 4.8 .

Locker No. → Step No. ↓	1	2	3	4	5	6	7	8	9	10	Lockers whose position is changed
	C	C	C	C	C	C	C	C	C	C	Initial position
1	O	O	O	O	O	O	O	O	O	O	1 to 10
2		C		C		C		C		C	2, 4, 6, 8, 10
3			C			O			C		3, 6, 9
4				O				O			4, 8
5					C					O	5, 10
6						C					6
7							C				7
8								C			8
9									O		9
10										C	10
No. of times locker position is changed	1	2	2	3	2	4	2	4	3	4	
Final position of lockers	O	C	C	O	C	C	C	C	O	C	

It is obvious that further steps say 11th, 12th etc. will not change the final position of the first 10 lockers as the 11th step will change the position of 11th, 22nd, 33rd lockers etc. So, the position reflected in Table 4.8 is final and complete for the first ten lockers. For ten lockers, it can be seen from the last row of Table 4.8 that lockers no. 1, 4 and 9 are finally left open and all other lockers are closed. Note that 1, 4 and 9 are perfect squares.

From the second last row of Table 4.8, it can be seen that if the position of the locker is changed an odd number of times say 1, 3 etc., the locker is finally left open else closed because the sequence of the position of lockers is open, close, open, close, open etc.

Why are only square numbered lockers finally left open? Consider a number say 6. There are four different divisors of the number 6 i.e., 1, 2, 3 and 6.

Now consider a perfect square number say 9. There are three different divisors of 9 i.e., 1, 3 and 9. We can see that except for perfect squares, all other numbers have an even number of divisors which comes in pairs such as

$$6 = 1 \times 6 = 2 \times 3, 15 = 1 \times 15 = 3 \times 5, 28 = 1 \times 28 = 2 \times 14 = 4 \times 7$$

and so on. The perfect squares have an odd number of divisors where two divisors are the same in one pair such as

$$9 = 1 \times 9 = 3 \times 3, 25 = 1 \times 25 = 5 \times 5 \text{ etc.}$$

The position of the locker is changed n number of times, where n is the number of divisors.

Since perfect squares have an odd number of divisors, the position of square numbered lockers is changed an odd number of times so they remain finally opened.

The concept explained above can be extended to any number of lockers.

In the puzzle, the number of lockers is 99.

The perfect squares up to 99 are 1, 4, 9, 16, 25, 36, 49, 64 and 81 i.e., nine lockers finally remain opened.

Comments: For N number of lockers, the number of lockers finally remain opened is equal to the integer value of $\sqrt{N}$.

If N = 300, $\sqrt{N}$ = 17.32, so for 300 lockers, 17 lockers numbered as $1^2, 2^2, 3^2, 4^2$ ... and 17^2 finally remain opened.

19. Weights in Increasing Order

Five items of different weights can be placed in increasing or decreasing order in seven weighings as given below:

Let the five silver items of different weights be designated as A, B, C, D, and E.

1. Weigh A against B, let A > B (i.e., A is heavier than B)
2. Weigh C against D, let C > D
3. Weigh heavier of items obtained in step (1) and step (2) i.e. A and C, let C > A. So, after three weighings, we have

$$B < A < C \tag{i}$$

$$C > D \tag{ii}$$

4. Now rank the item E by properly placing it in (i) i.e., $B < A < C$. For this weigh E against A

a) If $E > A$, weigh E against C

   (I)   If $E > C$, we get $B < A < C < E$ (iii)

   (II)   If $E < C$, we get $B < A < E < C$ (iv)

b) If $E < A$, weigh E against B

   (I)   If $E > B$, we get $B < E < A < C$ (v)

   (II)   If $E < B$, we get $E < B < A < C$ (vi)

After five weighings, we get four cases given by (iii), (iv), (v) and (vi) with additional condition given by (ii).

5. From (ii), it is already known that $C > D$, so we have to rank item D by properly placing it in (iii), (iv), (v) and (vi)

From (iii), $B < A < C < E$
Since $C > D$ from (ii), weigh D against B.
If $D < B$, we get $D < B < A < C < E$.
So, in this case, we obtain the objective in six weighings.
If $D > B$, then weigh D against A
If $D < A$ we get $B < D < A < C < E$
If $D > A$ we get $B < A < D < C < E$
So, in seven weighings, we obtain the solution.
From (iv), $B < A < E < C$
Since $C > D$ from (ii), weigh D against A.
If $D < A$ weigh D against B,
If $D < B$ we get the solution as $D < B < A < E < C$
If $D > B$ we get the solution as $B < D < A < E < C$
If $D > A$, weigh D against E,
If $D < E$ we get $B < A < D < E < C$
If $D > E$ we get $B < A < E < D < C$

So, in seven weighings, we obtained the solution. Similarly, for the remaining two cases, we can obtain the solution in seven weighings.

Comment: It is obvious that for two items, only one weighing is sufficient. Three items will require three weighings. Similarly, four items can be dealt with in five weighings. With five items, we have seen that seven weighings are required. The minimum weighings required for 6, 7, 8, 9, 10 and 11 items are 10, 13, 16, 19, 22 and 26 respectively.

20. Grant Distribution Among Districts

Let's tabulate the details of grants available and share of districts as shown in Table 4.9.

Table 4.9 .

District	Available grant (%)	% share of district
A1	100	4
A2	$100 - 4 = 96$	8% of 96 = 7.68
A3	$96 - 7.68 = 88.32$	12% of 88.32 = 10.5984% and so on

As per the above technique, it becomes complicated to calculate % share of all districts. So, let's solve the puzzle in a different way as follows:

The nth district is denoted by A_n and this district gets 4n% of the available grant at his turn. So, first district A_1 gets 4% of the available grant i.e., 100% at the beginning.

Let the remaining grant (available) at the turn of the nth district is x.

So, the share of the nth district = 4nx/100

After nth district gets its share, the remaining grant = $x - 4nx/100$

So (n + 1)th district gets $\frac{4(n+1)}{100}$ × remaining grant

$= \frac{4(n+1)}{100} \times (x - \frac{4nx}{100})$

The share of (n + 1)th district is more than nth district, if

$\frac{4(n+1)}{100} \times (x - \frac{4nx}{100}) > \frac{4nx}{100}$

$\Rightarrow$ (n + 1) (100 − 4n) > 100n

$\Rightarrow$ 4 (25 − n² − n) > 0

So, if $25 - n^2 - n > 0$, then (n + 1)th district will get higher share than nth district.

This is true for n = 1, 2, 3 and 4.

So, for n + 1 i.e., 4 + 1 = 5th district i.e., A_5 share is more than 4th district i.e., A_4. Hence fifth district A_5 gets the largest share of the grant.

Comments: As the grant is distributed, the remaining grant goes on reducing but the percentage share of districts goes on increasing. At a certain stage, the balance amount i.e., the remaining grant becomes so small that even with a larger percentage of share, the district gets a lesser share than the district which received the larger share earlier. So, the objective is to find the district which gets the largest share.

If there are n districts and each district gets 100/n percentage of available grant as a share, then the district which gets the largest share is $\sqrt{n}$.

So, for 25 districts, with each district getting $100/25 = 4\%$ of available grant as a share, the fifth district gets the largest share as $\sqrt{25} = 5$.

It will be an interesting exercise to find the district with the largest share, if there are 20 districts with each district getting 5% of the available grant as a share. In this case, you will find that the 4th and 5th district gets equal share which is the largest share also. It is interesting to know, why this is so.

21. Trucks to Transfer Payload

The main objective is to minimize the consumption of fuel. With one truck having full fuel tank, a maximum distance of 300 km can be covered to deliver the full payload.

With two trucks, one can go up to a certain distance x and then drop one truck after transferring its remaining fuel and payload to another truck, which can cover a distance of 300 km (if its fuel tank is full) to deliver the full payload. So, the distance x can be determined by equalizing the fuel spent from the truck (which needs to proceed ahead) by the fuel remaining in the truck to be dropped.

Since fuel spent from each truck after travelling x km. is equal to x km and fuel left in each truck after travelling x km is $(300 - x)$ km. So

$x = 300 - x \Rightarrow x = 150$ km.

So, one truck can be dropped at 150 km and the other truck, which now has a full fuel tank can cover 300 km more distance. So, with two trucks, maximum distance covered $= 150 + 300 = 450$ km.

Now let's consider 5 trucks. As explained above, we have to compute the distance x_1, x_2, x_3 and x_4 where trucks can be dropped one by one and their payload along with remaining fuel can be utilized optimally by recouping the fuel of other trucks going ahead.

Let the first truck is dropped at a distance of x_1 km. After travelling x_1 km, fuel consumed by every truck $= x_1$ km.

Fuel remaining in every truck $= (300 - x_1)$ km.

Now remaining fuel of the first truck (proposed to be dropped) must be equal to the fuel consumed by the remaining four trucks, so that fuel of the remaining four trucks can be fully recouped. Hence

$(300 - x_1) = 4x_1$

$\Rightarrow x_1 = 60$ km.

So, at the end of 60 km, the first truck is dropped and the remaining four trucks proceed further with the full fuel tank.

Let second truck is proposed to be dropped after travelling a further distance of x_2 km.

With the same logic as explained earlier,

Fuel remaining in second truck = Total fuel spent from three other trucks.

$$\Rightarrow (300 - x_2) = 3x_2$$

$$\Rightarrow x_2 = 75 \text{ km.}$$

Similarly, we can compute the distance $x_3 = 100$ km and $x_4 = 150$ km, where the third and fourth truck will be dropped. The fifth and final truck will cover the distance $x_5 = 300$ km with a full fuel tank to deliver the full payload at the end. So total distance covered to reach for delivery of full payload = 60 + 75 + 100 + 150 + 300 = 685 km.

Note that every time a truck is dropped, its payload is also transferred to the remaining trucks proceeding ahead.

Comments: We can generalize the puzzle for n trucks with full fuel capacity = C km.

As discussed earlier, the distance to be covered before each truck is dropped is computed based on the concept that "Fuel remaining in one truck at a distance x_1 (before dropping it) must be equal to the total fuel consumed by all other trucks going ahead", So

$$C - x_1 = (n - 1)x_1$$

$$\Rightarrow x_1 = C/n$$

Similarly for the second truck to be dropped at x_2,

$$C - x_2 = (n - 2)x_2$$

$$\Rightarrow x_2 = C/(n - 1)$$

For $(n - 1)$th truck to be dropped at x_{n-1},

$$C - x_{n-1} = [n - (n - 1)]x_{n-1}$$

$$\Rightarrow x_{n-1} = C/2$$

Final truck i.e., nth truck now with full fuel tank and full payload can cover a distance equal to C. So, the total distance covered to reach for delivering full payload is

$$= C + \frac{C}{2} + \cdots \frac{C}{n-1} + \frac{C}{n} \tag{i}$$

$$= C\left[1 + 1/2 + 1/3 + 1/4 + \cdots \frac{1}{n-1} + \frac{1}{n}\right] \cdots$$

Equation (i) is a general equation for n trucks with full fuel capacity as C km. For n = 5 trucks with C = 300 km as given in the puzzle, total distance

$$= 300(1 + 1/2 + 1/3 + 1/4 + 1/5)$$
$$= 685 \text{ km}.$$

It is interesting to note that the quantity given in bracket of equation (i) is the sum of a harmonic series i.e., $H_n = 1 + 1/2 + 1/3 + 1/4 + \cdots + 1/n$.

The sum of the harmonic series up to n terms can be computed from the formula given below:

$$H_n = S_n/n!$$

Where, $S_1 = 1$ and $S_n = n S_{n-1} + (n - 1)!$

22. Inverting an Equilateral Triangle

To minimize the number of coins required to move to invert the triangle, we have to look for a pattern or shape enclosing the maximum possible number of coins that do not change by inverting the triangle.

A key observation to Fig. 4.8 indicates that a hexagon enclosing 10 coins is symmetrical to both figures and does not change by inverting the triangle.

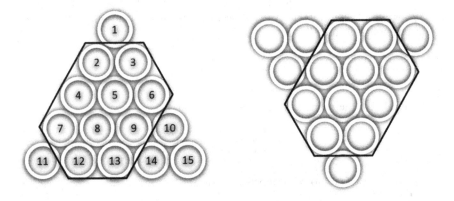

Fig. 4.8 .

So, these ten coins can be retained in their existing position. Only the remaining five coins need to be moved to invert the triangle.

Comments: If three coins are placed in a triangle form, only 1 coin is required to be moved to invert it. An equilateral triangle with 6 coins can be inverted by moving only 2 coins after retaining the symmetrical shape enclosing 4 coins as shown in Fig. 4.9.

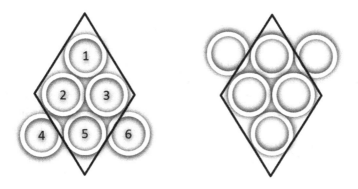

Fig. 4.9 .

Similarly, an equilateral triangle with 10 coins can be inverted by moving only 3 coins after retaining the symmetrical shape enclosing 7 coins as shown in Fig. 4.10.

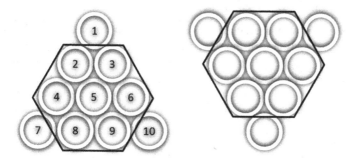

Fig. 4.10 .

If the equilateral triangle consists of N coins, then the minimum number of coins that need to be moved to invert (i.e., make an upside down) the triangle is equal to the integer value of $(\frac{N}{3})$.

Note that N is a triangular number that is equal to $\frac{n(n+1)}{2}$ where n = 1, 2, 3, 4... etc. So, N = 1, 3, 6, 10, 15, 21, ...

23. Largest Product from Distinct Numbers

Since the numbers are distinct positive integers, so the sum of 100 can be obtained either by considering fewer large numbers or many small numbers.

It is well known that if the sum of two numbers is 100, the largest product obtained by dividing 100 into two equal parts and then multiplying them is $(\frac{100}{2})(\frac{100}{2}) = 50^2$.

However, if the sum of four numbers is 100, then the largest product is $(\frac{100}{4})^4 = 25^4 > 50^2$.

This indicates that many small numbers must be considered for a given sum to obtain the largest product.

Obviously 1 is avoided as it does not help in increasing the product, though sum is increased by 1. So, starting from 2, we go on adding consecutive numbers and find that $2 + 3 + 4 + 5 + 6 + 7 + 8 + 9 + 10 + 11 + 12 + 13 = 90$.

The addition of the next number will make the sum more than 100 (i.e., $90 + 14 = 104$). So, to limit the sum to 100, we have to omit one number and add the next number i.e., 14. Obviously, number 4 is to be omitted as $90 + 14 - 4 = 100$.

Hence the smallest set of distinct positive integers > 1 whose sum is 100 is 2, 3, 5, 6, 7, 8, 9, 10, 11, 12, 13 and 14.

Product of $2 \times 3 \times 5 \times 6 \times 7 \times 8 \times 9 \times 10 \times 11 \times 12 \times 13 \times 14 = \frac{14!}{4} = 21794572800$.

This is the largest possible product, which can be obtained from distinct positive integers, whose sum is 100.

24. Gold Chain for Daily Payment

(i) As given, with one cut in between a chain, three pieces are obtained i.e., one piece of one cut link, one piece of say length a (i.e., a links) and one piece of say length b (i.e., b links). With these three pieces, maximum lengths possible are seven i.e., 1, a, b, a + b, a + 1, b + 1 and a + b + 1. So, one cut cannot cover the 23 lengths required for 23 days. Let us try cuts in two links, by which five pieces can be obtained as shown in Fig. 4.11.

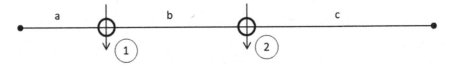

Fig. 4.11 .

One link cut 1, one link cut 2, a-link chain left of cut 1, b-link chain between cut 1 and cut 2, c-link chain right of cut 2.

With two pieces of one link each, two days payment can be done. For the third day, an additional link is required. Since 2 links already paid are available with the flat owner, if a 3-link chain is available and given to the flat owner, he can return two pieces of one link available with him. So, the first cut shall be made in the fourth link, so that a chain of 3-link (i.e., a = 3) is obtained.

With the availability of a 3-link chain and two single links, 5 days can be paid as

$$1, \ 1, \ 3\text{--}1\text{--}1, \ 1, \ 1$$

For the 6th day payment, the second cut needs to be made leaving 6 links after the first cut i.e., at the 11th link (So b = 6, and c = 12).

So, with 2 cuts i.e., at the 4th link and 11th link, the pieces obtained are two pieces of one link each, one piece of 3 links, one piece of 6 links and one piece of 12 links.

With these 5 pieces i.e., 1, 1, 3, 6 and 12, it is possible to make set of all integral lengths from 1 to 23 and day wise payment can be made as follows:

1, 1, 3–1–1, 1, 1, 6–3–1–1, 1, 1, 3–1–1, 1, 1, 12–6–3–1–1, 1, 1, 3–1–1, 1, 1, 6–3–1–1, 1, 1, 3–1–1, 1 and 1.

Note that 6–3–1–1 means, he gives 6 link chain and get back 3 link chain and two single links from the flat owner.

So, with two cuts i.e., at the 4th link and 11th link in 23-link chain, 23 days payment at the rate of one link per day can be made.

(ii) With 24-link chain, payment on the 24th day can be made by giving 24-link chain and taking back 23 links (i.e., 5 pieces of 23 links as given in (i) above) from the flat owner. With these 5 pieces of 23 links, payment can be made at the rate of one link per day for 23 days as already explained in (i) above. So, no cuts are required in 24-link gold chain.

Comments: We have seen that for a 23-link chain, two cuts are sufficient to make a set of all integral links from 1 to 23. Similarly, for a 7-link chain, one cut is sufficient at the 3rd link, which gives three pieces i.e., one link, 2 link chain and 4 link chain. Payment for 7 days, at the rate of one link per day, can be made as

$$1, \quad 2-1, \quad 1, \quad 4-3, \quad 1, \quad 2-1 \text{ and } 1.$$

This can be generalized to n cuts, which are sufficient for the N link chain to make a set of all integral links from 1 to N as given in Table 4.10.

Table 4.10 .

Number of cuts n	Number of pieces	Maximum number of links N
1	3	7 (1 + 2 + 4)
2	5	23 (1 + 1 + 3 + 6 + 12)
3	7	63 (1 + 1 + 1 + 4 + 8 + 16 + 32)
4	9	159 (1 + 1 + 1 + 1 + 5 + 10 + 20 + 40 + 80)
5	11	383 (1 + 1 + 1 + 1 + 1 + 6 + 12 + 24 + 48 + 96 + 192)
n	2n + 1	$n + (n + 1) [1 + 2 + 2^2 + 2^3 + \cdots + 2^n]$ $= n + (n + 1) (2^{n+1} - 1) = (n + 1) 2^{n+1} - 1$

It can be seen that 4 cuts i.e., at 6th, 17th, 38th and 79th are sufficient for 159-link chain.

Let us see how the formula for n cuts has been arrived at in Table 4.10.

For n cuts, n single links are obtained for n days. So for (n + 1)th day, we need a non-cut chain of (n + 1) links and this is the first non-cut chain with (n + 1) links. After that number of links doubles the previous one in every non-cut chain, this forms a geometric progression i.e.

$$n + (n + 1) + 2(n + 1) + 4(n + 1) + \cdots$$

This simplifies to $(n + 1) \, 2^{n+1} - 1$ as given in Table 4.10. Note that number of non-cut chains is $n + 1$.

25. **Age of Two Liars**

Let the age of liar A is X and the age of liar B is Y. Since all the statements of the liar A and liar B are wrong, we can conclude the following.

(i)   Liar A says that his age is certainly not more than 52. This implies that $X > 52$ is a correct statement.

(ii) Liar B says that he is 45 years old. This implies that Y cannot be equal to 45. Liar B also says that Liar A is at least 10 years older than him i.e. $(X - Y) \geq 10$. This implies that $(X - Y) < 10$ is a correct statement.

(iii) Liar A says that the age of liar B is more than 45 i.e., $Y > 45$. This implies that $Y \leq 45$ is a correct statement.

So, in a nutshell, we have

$X > 52, X - Y < 10, Y \neq 45$ and $Y \leq 45$

$\Rightarrow X > 52, X - Y < 10$ and $Y < 45$

To satisfy the above, the minimum possible value of $X = 53$ and the maximum possible value of $Y = 44$.

The difference $X - Y = 53 - 44 = 9 < 10$.

This satisfies all conditions. Taking any other value of X and Y will not satisfy the condition $X - Y < 10$. Hence the age of Liar A = 53 years and the age of Liar B = 44 years.

26. **Three Sundays on Prime Numbered Dates**

The prime numbered dates in a month can be 2, 3, 5, 7, 11, 13, 17, 19, 23, 29 and 31 only. The first Sunday (or any weekday) can fall on any day from the 1st to the 7th day of a month. If the first Sunday (or any weekday) falls on the 4th to 7th day of a month, there can be only four Sundays (or any weekday) in that month, out of which two Sundays will fall on odd days and two on even days ≥ 4. In such a case, only two Sundays can fall on prime numbered dates. Hence for three Sundays to fall on prime numbered dates, the first Sunday must fall on the 1st to the 3rd day of the month. Accordingly, possible days of the month on which Sundays can fall are:

$$1, 8, 15, 22, 29$$
$$2, 9, 16, 23, 20$$
$$3, 10, 17, 24, 31$$

It can be seen that only the third combination contains three prime numbered dates i.e., 3, 17 and 31, on which three Sundays can fall. So, the first Sunday must fall on the 3rd of the month, thus 1st day of the month will fall on Friday.

Since the birthday of Om Prakash falls on the 1st day of the month of the second quarter of a calendar year, the month containing 31 days is only 'May' in the second quarter of any calendar year (April and June contain 30 days). So, the birthday of Om Prakash will fall on Friday, 1st May.

Comments: If instead of prime numbered dates, three Sundays fall on even-numbered dates of that month, then it is possible only if five Sundays fall in that month. So, the corresponding even-numbered dates can only be 2, 16 and 30th. So, if the 2nd day is Sunday, then the 1st day of the month will fall on Saturday.

27. Squares from Adjacent Numbers

To find the solution, the first question arises is, where to start? This is not difficult.

We can see that the possible perfect squares, which can be the sum of adjacent numbers, are 4, 9, 16 and 25, as the sum of the largest two numbers is $15 + 14 = 29$, which is less than 36.

Now we can work out the possible pairs of adjacent numbers, which sum to these squares as given in Table 4.11.

Table 4.11 .

Square	Possible pairs of numbers sum
4	1 + 3
9	**1 + 8**, 2 + 7, 3 + 6, 4 + 5
16	1 + 15, 2 + 14, 3 + 13, 4 + 12, 5 + 11, 6 + 10, **7 + 9**
25	10 + 15, 11 + 14, 12 + 13

From Table 4.11, it can be seen that number 8 is adjacent to 1 only and number 9 is adjacent to 7 only. So, for the solution to exist, these two numbers i.e., 9 and 8 must be placed at the ends.

Let's start with the number 9 and put adjacent numbers 1 to 15 such that the sum of two adjacent numbers is a perfect square.

$9-7-2-14-11-5-4-12-13-3$

The above is the only sequence possible from 9 up to 3.

From number 3 onwards, there are two possibilities.

(i) $3-1-8$
(ii) $3-6-10-15-1-8$

Since case (i) does not cover all the balance numbers from 1 to 15, so only case (ii) is appropriate. The complete solution, therefore, is represented as,

$9-7-2-14-11-5-4-12-13-3-6-10-15-1-8$

Based on similar logic, we can start with the number 8 and complete the sequence ending with 9 i.e., the reverse of the above sequence. These are the only solutions.

Comments: In case the puzzle requires the numbers from 1 to 16 or 1 to 17, such that the sum of two adjacent numbers is a perfect square, then 16 can be added adjacent to 9 to complete the solution for N = 16 and then 17 can be added adjacent to number 8 to complete the solution for N = 17.

There are no solutions for N = 18 to N = 22.

Another variant of the puzzle requires the numbers from 1 to 32 to be placed in a circle, such that the sum of every two adjacent numbers is a perfect square. Proceeding similarly, it can be observed that the largest square must be less than or equal to 32 + 31 = 63. So possible squares, from the sum of the two adjacent numbers, are 4, 9, 16, 25, 36 and 49.

For numbers from 25 to 32, there are only two possible adjacent numbers that can make the sum a perfect square which is 36 or 49. These combinations are: (25 + 11, 25 + 24), (26 + 10, 26 + 23), (27 + 9, 27 + 22), (28 + 8, 28 + 21), (29 + 7, 29 + 20), (30 + 6, 30 + 19), (31 + 5, 31 + 18) and (32 + 4, 32 + 17). Keeping in view the above numbers, the remaining numbers can be appropriately placed to arrive at the final solution given in Fig. 4.12.

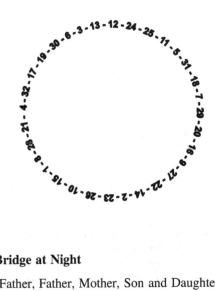

Fig. 4.12 .

28. **Crossing the Bridge at Night**

(a) Let Grand Father, Father, Mother, Son and Daughter be denoted by G, F, M, S and D respectively. The solution is given in Table 4.12.

Table 4.12 .

Crossing No.	Left side	Bridge	Right side	Time (minutes)
	G, F, M, S, D			
1	G, F, M	$\xrightarrow{S,D}$	S, D	2
2	G, F, M, S	$\xleftarrow{S}$	D	1
3	G, M	$\xrightarrow{S,F}$	D, S, F	6
4	G, M, S	$\xleftarrow{S}$	D, F	1
5	S	$\xrightarrow{G,M}$	D, F, G, M	11
6	S, D	$\xleftarrow{D}$	F, G, M	2
7	–	$\xrightarrow{S,D}$	G, M, F, S, D	2

So, the minimum time to cross the bridge
$= 2 + 1 + 6 + 1 + 11 + 2 + 2 = 25$ minutes
There are other solutions also.
The above details can be put in a pictorial form as given in Fig. 4.13.

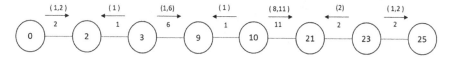

Fig. 4.13 .

The number below the arrow denotes the crossing time of a slower person on the bridge and the number above the arrow denotes the crossing time of a person/persons travelling on the bridge. The direction of the arrow denotes the direction of movement of people. Numbers in the circle denote the cumulative time. It can be seen that the minimum time required for the family to cross the bridge is 25 minutes, which is given in the last circle.

(b) The crossing time of the daughter is changed from 2 minutes to 5 minutes. The minimum time now required to cross the bridge changes.

If we proceed in the way explained in (a) above, the minimum time required is 34 minutes as per the details given in the pictorial form in Fig. 4.14.

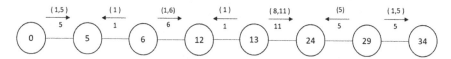

Fig. 4.14 .

The above timings of 34 minutes can be improved as detailed in pictorial form in Fig. 4.15.

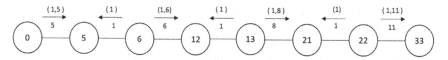

Fig. 4.15 .

It can be seen that by making suitable changes in the last three steps, the time has been reduced from 34 minutes to 33 minutes. Let us analyse these two cases for the last three steps.
Case I: Here two slowest persons cross, then the second fastest return and finally the two fastest cross. Time taken is 34 minutes.

Case II: Here fastest and second slowest cross, then the fastest return and finally the fastest and slowest cross. Time taken is 33 minutes.

So, let's now discuss the methodology to be adopted for solving such puzzles in general.

If there are N people with crossing times as $t_1 < t_2 < t_3 \cdots < t_{N-1} < t_N$, where t_1 is fastest and t_N is slowest.

(i)   Except t_N and t_{N-1}, all others cross the bridge in pairs as (t_1, t_x) where x is from 2 to N − 2 and t_1 returns.

(ii) After completing step (i) above, we are left with t_1, t_{N-1} and t_N which are yet to cross the bridge, their crossing depends on the values of t_1, t_{N-1} and t_2 as follows:

Case I: $t_1 + t_{N-1} \geq 2t_2$

In this case, the two slowest persons (t_N, t_{N-1}) cross the bridge, the second fastest person t_2 returns and finally fastest pair (t_1, t_2) crosses.

Case II: $t_1 + t_{N-1} < 2t_2$

In this case pair (t_1, t_{N-1}) cross the bridge, fastest person t_1 return and finally (t_1, t_N) cross the bridge.

It can be seen that, the minimum time required to cross the bridge in case I is

= $t_1 \times (N - 3)$ + Sum of timings from t_2 to t_{N-2} + $2t_2 + t_N$

and for case II, the minimum time required is

= $t_1 \times (N - 3)$ + Sum of timings from t_2 to t_{N-2} + $t_1 + t_{N-1} + t_N$

In the given puzzle a) we have N = 5 and $t_1 = 1$, $t_2 = 2$, $t_3 = 6$, $t_4 = 8$ and $t_5 = 11$

Since $t_1 + t_{N-1} = 1 + 8 = 9 \geq 2t_2$, so case I is applicable and minimum time required

= $1 \times 2 + 2 + 6 + 2 \times 2 + 11 = 25$ minutes.

In puzzle b) we have N = 5 and $t_1 = 1$, $t_2 = 5$, $t_3 = 6$, $t_4 = 8$ and $t_5 = 11$

Since $t_1 + t_{N-1} = 1 + 8 = 9 < 2t_2$, so case II is applicable and minimum time required

= $1 \times 2 + 5 + 6 + 1 + 8 + 11 = 33$ minutes.

Comments: For N persons, the number of crossings involved is N − 2 single crossings (Returning) and N − 1 double crossings. So total number of crossings involved is N − 2 + N − 1 = 2N − 3.

For N = 5 persons, it can be seen that $2 \times 5 - 3 = 7$ crossings are required.

Let's now consider the case where instead of 2 persons, a maximum of 3 persons are allowed to cross the bridge at a time.

Consider 10 persons with crossing times (in minutes) as 1, 2, 3, 4, 5, 6, 7, 8, 9 and 10, who want to cross the bridge, where a maximum of 3 persons are permitted at one time. Other conditions remain the same as in the original puzzle.

The solution in brief for the minimum time required to cross the bridge is given in Fig. 4.16.

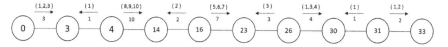

Fig. 4.16 .

So, the minimum time required by 10 persons to cross the bridge = 33 minutes.

29. Distinct Digit Milometer Reading

Let the car milometer reading (in km) at the beginning of the journey is AAAAA.

Let the five distinct digit milometer reading (in km) is ABCDE. First digit A has been kept the same as the difference between ABCDE and AAAAA is required to be minimum.

For the minimum value of ABCDE − AAAAA, BCDE − AAAA must be minimum, so AAAA has to be maximum. Since A cannot be 9 as in that case ABCDE − AAAAA will be negative, so let A = 8.

The starting reading AAAAA = 88888 implied B = 9. For the smallest difference between ABCDE and AAAAA, where A = 8, B = 9, the value of CDE must be taken as 012.

So, the shortest distance required to be travelled = 89012 − 88888 = 124 km.

30. Three Hands of a Clock

Let's first find the times during 12 hours when hour hand and minute hand overlap.

Let H is the angle in degrees which hour hand makes from 12 o'clock position after time t hours. Let M is the angle in degrees which minute hand makes from 12 o'clock position after time t hours. Refer to Fig. 4.17.

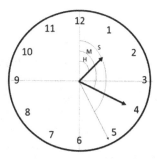

Fig. 4.17 .

Since the minute hand completes one revolution i.e., 360° in one hour and the hour hand completes 1/12th revolution i.e., 360/12 = 30° in one hour, so in t hours, the angle made by the hour hand from 12 o'clock position i.e., H = 30t.

Similarly, the angle made by minute hand from 12 o'clock position in t hours i.e., M = 360t

Hence the angle between the minute hand and hour hand = 360t − 30t = 330t

For the minute hand and hour hand to overlap, the angle between the minute hand and hour hand must be zero or a multiple of 360° (i.e., an integer multiple of one revolution).

So 330t = 360n, where n = 0, 1, 2, 3 ...

$$\Rightarrow t = \frac{12n}{11} \qquad\qquad (i)$$

Hence minute hand and hour hand overlaps at

$$t = 0, \frac{12}{11}, \frac{24}{11}, \frac{36}{11}, \dots \frac{120}{11} \text{ hours.}$$

Now let's find the times when hour hand and second hand overlaps.

Let angle made by second hand is S degrees from 12 o'clock position in t hours. Since second hand makes 60 revolutions per hour, so

S = 360t × 60 = 21600t

For the second hand and hour hand to overlap, the angle between the second and hour hand must be zero or a multiple of 360° (i.e., integer multiple of one revolution), so

S − H = 21600t − 30t = 360 m, where m = 0, 1, 2 ...

$$\Rightarrow t = \frac{360m}{21570} = \frac{12m}{719} \qquad\qquad (ii)$$

If all three hands i.e., hour, minute and second hand are to overlap at time t hours after 12 o'clock position, then from (i) and (ii) = $\frac{12n}{11} = \frac{12m}{719}$

$$\Rightarrow \frac{n}{m} = \frac{11}{719} \qquad\qquad (iii)$$

Since 11 and 719 have no common factor (both being primes), m and n being integers, Eq. (iii) cannot be satisfied except when n is a multiple of 11 and m is a multiple of 719, which is possible only at 12 hours or it's multiple, hence three hands of a clock can never overlap except at 12 o'clock position. (i.e., noon and midnight).

Comments: The minute and second hand overlap 59 times in 12 hours time as

S − M = 21600t − 360t = 360p, where p = 0, 1, 2 ...

$$\Rightarrow t = \frac{360p}{21240} = \frac{p}{59}$$

Since 59 is also a prime number, hence cannot have a common factor with 11 or 719. This gives additional proof that all the three hands of a clock cannot overlap except at 12 hours.

Let's now examine the times other than 12 o'clock, when the hour hand, minute hand and second hand come closest together. Closest means two hands of a clock overlap and the third hand is at a minimum distance away.

Consider the overlapping times of the two slowest hands of a clock i.e., hour hand and minute hand as given below:

$$t = 0, 1\frac{1}{11}, 2\frac{2}{11}, 3\frac{3}{11}, 4\frac{4}{11}, 5\frac{5}{11}, 6\frac{6}{11}, 7\frac{7}{11}, 8\frac{8}{11}, 9\frac{9}{11} \text{ and } 10\frac{10}{11} \text{ hours.}$$

Examine the position of the second hand at these timings and find the times when the second hand is closest to them.

At $t = 1\frac{1}{11} = 1$ hours $\frac{60}{11}$ minutes = 1 hours $5\frac{5}{11}$ minutes, seconds hand is away by (5/11)th of one revolution from 12 o'clock as against (1/11)th of minutes hand.

At $t = 2\frac{2}{11} = 2$ hours $10\frac{10}{11}$ minutes, seconds hand is away by (10/11)th of one revolution form 12 o'clock position as against (2/11)th of minutes hand.

At $t = 3\frac{3}{11} = 3$ hours $16\frac{4}{11}$ minutes, seconds hand is away by (4/11)th of one revolution from 12 o'clock position as against (3/11)th of minutes hand.

Proceeding similarly, one can find that the seconds hand is closest when the hour hand and minute hand overlap at $t = 3\frac{3}{11}$ hours i.e., 3 hours $16\frac{4}{11}$ minutes and its mirror image i.e., at $t = 8\frac{8}{11} = 8$ hours $43\frac{7}{11}$ minutes.

We can obtain the exact solution (i.e., when three hands come closest) by shifting the second hand till it coincides with the hour hand, the minute hand will also shift back slightly.

As seen earlier, the hour hand and second hand overlap 719 times in 12 hours. The times at which these two hands overlap is given by Eq. (ii) as $t = \frac{12m}{719}$ where $m = 0, 1, 2 \ldots 718$.

The overlap of hour hand and seconds hand, which comes closest to $t = 3\frac{3}{11}$ hours can be obtained by equating $t = \frac{12m}{719}$ and $t = 3\frac{3}{11}$, which gives $m = 196$ and the closest time is given by $t = \frac{12 \times 196}{719}$ hours = 3 hours 16 minutes $16\frac{256}{719}$ seconds and the corresponding symmetric solution is 12 hours – t = 8 hours 43 minutes $43\frac{463}{719}$ seconds.

31. Review Meeting Days

(i) The review meeting with HR, operations, business development, design and finance department is on every 2nd day, 3rd day, 4th day, 5th day and 6th day respectively.

LCM of 2, 3, 4, 5 and 6 = 60.

So, review meeting with all five departments will be on every 60th day from 1st January.

Since 2020 is a leap year, so February will have 29 days. Total number of days in January and February = 31 + 29 = 60 days. So, after 1st January, the 60th day will fall on 1st March.

Hence the Managing Director will again meet with directors of all departments on 1st March 2020.

(ii) The date of meeting with all the departments is 1st January and 1st March. To find the days, when there is no review meeting with any of the departments between 1st January and 1st March proceed as follows.

Write down all dates from 1st January to 1st March and strike out all the days of review meetings held by different departments i.e., strike out 1st January, 3rd January, 5th January etc. (every second day when a review meeting is held with the HR department). Similarly strike out 4th January, 7th January, 10th January, etc. (days of review meeting with operations department) and so on till all the dates are stroked out concerning the days of review meeting held with all departments. We are left with the following days when there are no review meetings.

January:	2, 8, 12, 14, 18, 20, 24 and 30
February:	1, 7, 11, 13, 17, 19, 23 and 29

Comments: A more appropriate solution to the second part of the puzzle i.e., to find days when there is no review meeting with any of the departments is as given below.

January 2 is the first day when there are no review meetings. The other days when there are no review meetings are:

a) p days after 1st January, where p is a prime number greater than 6 and less than 60. This is because p will not have any common factor with 60.

The primes p are 7, 11, 13, 17, 19, 23, 29, 31, 37, 41, 43, 47, 53 and 59. The corresponding days when there is no review meeting are

January:	8, 12, 14, 18, 20, 24, 30
February:	1, 7, 11, 13, 17, 23, 29

b) In addition to the above, there are more days that also do not have any common factor with 60, these are $p \times p_1$ days after 1st January, where p_1 and p are primes $p_1 \geq p$ and $p > 6$, also $pp_1 < 60$.

The corresponding values of $pp_1 = 7 \times 7, 7 \times 11 \dots$ and $11 \times 11, 11 \times 13$ etc.

One can see that except $pp_1 = 7 \times 7$, all other values are greater than 60.

So correspondingly 49 days after 1st January, there will be no review meeting, this will fall on 19th Feb. It can be verified that both methods give the same result but the second method is more appropriate being a general method.

32. Curious Olympic Rings

It can be seen that numbers in regions A, C, E, G and I are used only once, whereas numbers in regions B, D, F and H are used twice.

Sum of all numbers $= 1 + 2 + 3 + 4 + 5 + 6 + 7 + 8 + 9 = 45$

There are five rings and sum of numbers in each ring is equal to 14, so

$$B + D + F + H + 45 = 5 \times 14$$
$$\Rightarrow B + D + F + H = 25 \tag{i}$$

From the rings, it can be seen that

$$B + D + C = 14 \text{ and } F + H + G = 14$$
$$\Rightarrow B + D + C + F + H + G = 28 \tag{ii}$$

From Eqs. (i) and (ii)

$$C + G = 28 - 25 = 3 = 1 + 2$$

So, C and G take the values from 1 and 2.

Since A + B = 14, I + H = 14 and $14 = 9 + 5 = 8 + 6$ (without repetition of any digit),

So, A, B, I and H can only take values from 5, 6, 8 and 9.

The remaining digits are 3, 4, 7, which will be taken by the remaining letters E, D and F.

$$E + D + F = 14 = 3 + 4 + 7 \tag{iii}$$

So, D + F can be 3 + 4, 3 + 7 or 4 + 7 i.e., 7, 10 or 11 only.

From Eq. (i) B + D + F + H = 25

If D + F = 7 $\Rightarrow$ B + H = 25 − 7 = 18 (not possible).

If D + F = 10 $\Rightarrow$ B + H = 25 − 10 = 15 = 8 + 7 = 9 + 6 (Since 7 is already taken by E/D/F, so B + H = 9 + 6).

If D + F = 11 $\Rightarrow$ B + H = 25 − 11 = 14 = 9 + 5 = 8 + 6 (Not possible as 5, 9, 6 and 8 belongs to A, B, I and H only). Hence D + F = 10 = 3 + 7 and B + H = 15 = 6 + 9 only.

$\Rightarrow$ E = 4 (from Eq. iii).

So, B and H can take the values from 6 and 9.

D and F can take the values from 3 and 7.

A and I can take the values from 5 and 8.

C and G can take the values from 1 and 2.

E = 4.

With the combination of values found above, it is easy to find values of each unknown.

Consider A + B = 14 = 9 + 5 = 8 + 6

Let A = 9, So B = 5 (But B can only be 6 or 9), so A = 5 and B = 9.

With A = 5 and B = 9, I = 8 and H = 6

Now B + D + C = 14 $\Rightarrow$ D = 5 – C (as B = 9)
If C = 1, D = 4 (Not possible as E = 4), so C = 2 and D = 3, so G = 1
Consider F + H + G = 14 $\Rightarrow$ F = 7 (As H = 6 and G = 1).
Hence A = 5, B = 9, C = 2, D = 3, E = 4, F = 7, G = 1, H = 6, I = 8 as
shown in Fig. 4.18.

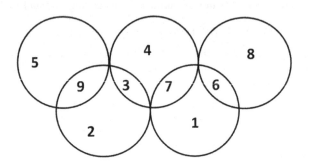

Fig. 4.18 .

If we start with A + B = 14 = 8 + 6, then we get other solution which is mirror
image of above solution i.e. A = 8, B = 6, C = 1, D = 7, E = 4, F = 3, G = 2,
H = 9 and I = 5 as shown in Fig. 4.19.

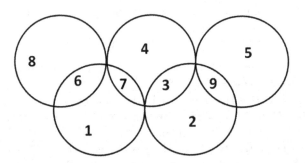

Fig. 4.19 .

33. Maximum Area of Plot

The key point to note is that in the case of three given sides, only one triangle
can be formed but the same is not true for the given four sides, where any
number of quadrilaterals is possible. The owner has decided to sell the plot with
sides of 5 m, 9 m, 13 m, 15 m and an area of 80 sqm for 4000000 rupees,
which gives a rate of 4000000/80 = 50000 rupees per sqm. With the given four
sides, it is possible to construct a quadrilateral with an area larger than 80 sqm
that is why the owner has given a second option with extra charges of 10%.
Before deciding upon the second option, the purchaser would like to know, the

largest possible area of a quadrilateral that can be enclosed with the given four sides i.e., 5 m, 9 m, 13 m and 15 m.

If the four sides are represented by a, b, c and d then the largest possible area of the quadrilateral is given by $A = \sqrt{(s-a)(s-b)(s-c)(s-d)}$

Where s is half the perimeter and is equal to $(a + b + c + d)/2$

Putting the values of a, b, c and d, we get $s = (5 + 9 + 13 + 15)/2 = 21$ and

$$A = \sqrt{(21-5)(21-9)(21-13)(21-15)}$$
$$= \sqrt{16 \times 12 \times 8 \times 6} = 96 \text{ sqm.}$$

If 10% extra cost is to be paid for 96 sqm, then total cost to be paid = 4400000 rupees.

So rate $= 4400000/96$

$= 45833.33$ rupees per sqm < 50000 rupees per sqm.

So, the purchaser will opt for the second option and would like to get 96 sqm of land for the plot at a total cost of 4400000 rupees.

Comments: Consider a quadrilateral ABCD, with sides as a, b, c and d as shown in Fig. 4.20.

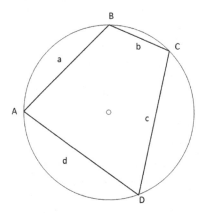

Fig. 4.20 .

When the sum of opposite angles is 180° i.e., Angle BAD + Angle BCD = 180, the area of quadrilateral will be maximum and is given by $A = \sqrt{(s-a)(s-b)(s-c)(s-d)}$, where s is half the perimeter.

If the four sides are so adjusted that corners (vertices) A, B, C and D of the quadrilateral lie on a circle, then this quadrilateral will give the maximum area for the given sides. Such a quadrilateral is known as a cyclic quadrilateral and in this case sum of opposite angles is 180°.

34. Building Location for Senior Citizen

Since the distances are not given, this indicates that there is no need to form equations.

(i) First consider that there are only two buildings (extreme) A and H where his friends are staying. Now consider any location X anywhere between A and H.

$$\underline{\qquad\overset{\displaystyle A\qquad\qquad X\qquad\qquad H}{}\qquad}$$

Sum of the distances of X from A and H = XA + XH = AH, which is fixed irrespective of the location of X. If X is at A, then AX = 0, XH = AH, so the total distance is again AH.

However, if X is located outside of A or outside of H, the sum of distances of X from A and H will be higher than AH as calculated above. This proves that any location between the two buildings A and H will give the smallest sum of distances from A and H.

Now consider the next two buildings where his friends are staying i.e., B and G. As explained earlier, the smallest sum of distances from any location X will be obtained if X is located anywhere between B and G (Note that in this case also X remains between A and H). So, if we consider four buildings A, H, B and G, then any location of X between B to G will give the smallest sum of distances from X to A, H, B and G.

Proceeding in this way and taking each successive pair of buildings in which, his friends stay, it can be seen that any location of X, anywhere from D to E will minimize the sum of distances from X to A, H, B, G, C, F, D and E.

Now from D to E, there is one other building d, so the senior citizen can select buildings D, d or E for his stay which will minimize the sum of distances from his building to all the other 8 buildings where his friends stay.

(ii) If his friends staying in building H left away, so now the remaining buildings are seven i.e. A, B, C, D, E, F and G whose sum of distances are to be minimized. Proceeding similarly as in (i) above, it can be seen that for A, G, B, F, C, E, the preferred location of building for a senior citizen can be anywhere from C to E for minimizing the distances to all buildings except D. To minimize the distance to building D also, he must locate in the building D only. So, in this case, he has only one choice i.e., building D for his stay which minimizes the sum of distances of all buildings where his friends stay.

Comments: There are two cases depending upon the number of buildings whether even or odd, whose sum of distances is to be minimized from the given location.

If the number of buildings is even, say 2n, then any location from nth building to (n + 1)th building (i.e. two middle buildings) will give the minimum sum of distance, for all 2n buildings.

If the number of buildings is odd, say 2n + 1, then location exactly at (n + 1)th building (i.e. central building) will give the minimum sum of distances from all (2n + 1) buildings.

35. Roll Number Dilemma

As per condition a), the roll number can be 44 or 48.

From condition b), if it is not a multiple of 5, then the roll number can be 51, 52, 53, 54, 56, 57 or 58. Since neither 44 nor 48 is a multiple of 5, as per condition b), the roll number cannot be a multiple of 4.

Since roll number cannot be multiple of 4, so it cannot be multiple of 8 also. So, from condition c), the possible roll numbers can be 61, 62, 63, 65, 66, 67 or 68. The only roll number which is a multiple of 5 is 65. Hence correct roll number of the student is 65, as in this case his roll number cannot be decided by condition b).

36. Relative Motion on Circular Track

Initially, the ratio of the speed of vehicle X and vehicle Y is 3/2. Let the speed of vehicle Y is 2V, so the speed of vehicle X will be 3V. Since both travel in opposite directions, their relative speed will be 2V + 3V = 5V. To make the puzzle simpler, divide the circular track into five equal segments namely AB, BC, CD, DE and EA, each measuring 20 km as shown in Fig. 4.21.

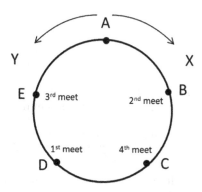

Fig. 4.21 .

When vehicle X completes 3 segments i.e., AB, BC and CD, the vehicle Y will complete 2 segments i.e., AE and ED, thus meeting vehicle X at D. At this meeting point, the two vehicles change their direction of travel and interchange their speed, so the speed of Y becomes 1.5 times the speed of X and this vehicle Y will travel towards E and vehicle X will travel towards C. When vehicle Y covers three segments from first meeting point D, vehicle X will cover two segments i.e., DC and CB so both vehicles will meet at B which is the point of the second meeting. At B, again the speed of vehicles gets interchanged and the direction of travel is reversed. Proceeding similarly, the point of the third meeting and fourth meeting will be at E and C respectively.

At the point of the fourth meeting i.e., C, vehicle X travel towards D at 1.5 times the speed of Y and Y will travel towards B, so they both meet again at A which is the point of the fifth meeting.

The details are summarized in Table 4.13.

Table 4.13 .

Point	Direction of vehicle X	Direction of vehicle Y	Speed of vehicle X	Speed of vehicle Y	Distance travelled by vehicle X (km)	Distance travelled by vehicle Y (km)
Start point A	Clockwise	Anti-clockwise	3V	2V	–	–
I meet point D	Anti-clockwise	Clockwise	2V	3V	60	40
II meet point B	Clockwise	Anti-clockwise	3V	2V	40	60
III meet point E	Anti-clockwise	Clockwise	2V	3V	60	40
IV meet point C	Clockwise	Anti-clockwise	3V	2V	40	60
V meet point A	–	–	–	–	60	40
					260 km	**240 km**

It can be seen from Table 4.13, that the total distance travelled by X and Y, when they meet again at A is 260 km and 240 km respectively.

Comments: Since both vehicles start at the same time and meet again at the same time, the time taken by both vehicles will be the same at all meeting points.

If the initial speed of vehicle Y is given, say 50 km/h, the time taken by both vehicles to meet again at A can be computed easily as under:

Initial speed of vehicle X = 1.5 × 50 = 75 km/h

Relative speed = 50 + 75 = 125 km/h

Total distance covered by both vehicle = 260 + 240 = 500 km

So, the time taken by vehicles to meet at A again = 500/125 = 4 hours

So, both the vehicles will meet at point A again at 7 AM + 4 hours = 11 AM

For such puzzles, the problem can be simplified, if the circular length of the track is divided into segments according to the relative speed of the vehicles. In the present case, the ratio of the speed of both vehicles at the starting point is 3:2, so it was decided to divide the circular track into 3 + 2 = 5 equal segments. If the ratio of speeds is m : n, then the circular length of the track is divided into m + n equal segments.

37. Race Competition

In the first round, teams A, B and C covers 400 m, 390 m and 375 m respectively in time t. So, their speeds are 400/t, 390/t and 375/t respectively. In the second round,

$$\text{time taken by any team} = \frac{\text{distance covered by the team}}{\text{speed of the team}}$$

So,

$$\text{Time taken by team A} = \frac{425}{400/t} = 1.0625\,t$$

$$\text{Time taken by team B} = \frac{415}{390/t} = 1.0641\,t$$

$$\text{Time taken by team C} = \frac{400}{375/t} = 1.066\,t$$

It can be seen that team A is the winner in the second round also, as it takes the least time to finish the race.

Comments: This puzzle can be solved without going into the computations shown above.

Let's first consider team A and team B.

In the first round, team A runs 400 m, at the same time team B runs 390 m. So, in the second round also (because of the constant same speed as in the first round), team A will cover 400 m at the same time, say t as team B covers 390 m. So, after time t, both team A and B meets. Since the speed of team A is more than team B, so the remaining distance (which is equal for both teams) will be covered in less time by team A as compared to team B.

Based on similar logic, team B will finish earlier than team C. So, team A will be the winner.

38. A Gem from Bhaskaracharya II's Work

Let the two numbers are x and y, so

$$x^2 + y^2 = z^3 \text{(A cube)} \qquad \text{(i)}$$

$$x^3 + y^3 = w^2 \text{(A square)} \qquad \text{(ii)}$$

Equation (ii) can be satisfied by $x = 1$ and $y = 2$, so let $x = a^2$ and $y = 2a^2$. Putting these values of x and y in Eq. (ii)

$$\left(a^2\right)^3 + \left(2a^2\right)^3 = 9a^6 = \left(3a^3\right)^2 \text{ which is a perfect square.}$$

Now let's put these values of x and y in Eq. (i)

$$\left(a^2\right)^2 + \left(2a^2\right)^2 = 5a^4$$

For $z^3 = 5a^4 = (5a) \times a^3$, 5a must be a cube.
To make it a cube, let $a = 5 \times 5r^3$ for any positive integer value of r. This gives $x = a^2 = 625r^6$ and $y = 2a^2 = 1250r^6$.
For $r = 1$, this gives the required first pair $(x, y) = (625, 1250)$, which satisfy the given conditions i.e.,
$$625^2 + 1250^2 = 25^3\left(5^2 + 10^2\right) = 25^3 \times 5^3 = 125^3 \text{ and}$$
$$625^3 + 1250^3 = 625^3(1 + 8) = 3^2 \times \left(25^3\right)^2 = 46875^2$$
Further solutions can be obtained by taking different values of r.

Comments: Instead of choosing the initial values of $x = a^2$, $y = 2a^2$, one can choose other values to satisfy Eq. (ii).
For example, $x = 11^2a^2$, $y = 37^2a^2$ as $11^3 + 37^3 = 228^2$, or $x = 56^2a^2$, $y = 65^2a^2$ as $56^3 + 65^3 = 671^2$. Author has named the pairs of such numbers (x, y) as Bhaskara pairs. If $x = y$, then these can be termed as Bhaskara twins. First few Bhaskara twins are (2, 2), (128, 128), (1458, 1458).
Similarly, we can find Bhaskara pairs, some of these are (1250, 625), (9774, 11611), (45162, 48862), (80000, 40000), (178245, 174284), etc.

39. Five Children and 517 Gold Coins

Let the least number of gold coins received by any child be n_1. So, the second-lowest number of gold coins, the third-lowest number of gold coins, the fourth-lowest number of gold coins and the fifth-lowest (i.e., highest) number of gold coins can be represented by n_1n_2, $n_1n_2n_3$, $n_1n_2n_3n_4$ and $n_1n_2n_3n_4n_5$ respectively, where n_1, n_2, n_3, n_4 and n_5 are positive integers.
Total number of gold coins = 517
$\Rightarrow n_1 + n_1n_2 + n_1n_2n_3 + n_1n_2n_3n_4 + n_1n_2n_3n_4n_5 = 517 = 11 \times 47$
$\Rightarrow n_1 (1 + n_2 + n_2n_3 + n_2n_3n_4 + n_2n_3n_4n_5) = 11 \times 47$
Since 11 and 47 both are prime numbers and $n_1 > 1$, so $n_1 = 11$.
So, $1 + n_2 + n_2n_3 + n_2n_3n_4 + n_2n_3n_4n_5 = 47$

$\Rightarrow n_2 (1 + n_3 + n_3n_4 + n_3n_4n_5) = 47 - 1 = 46 = 2 \times 23$

Since 2 and 23 both are prime numbers, so $n_2 = 2$.

Similarly, $1 + n_3 + n_3n_4 + n_3n_4n_5 = 23$

$\Rightarrow n_3 (1 + n_4 + n_4n_5) = 23 - 1 = 22 = 2 \times 11$

Since 2 and 11 both are prime numbers, so $n_3 = 2$.

Now $1 + n_4 + n_4n_5 = 11$

$\Rightarrow n_4 (1 + n_5) = 11 - 1 = 2 \times 5$

So $n_4 = 2$ and $n_5 = 4$.

Hence the number of gold coins received by five children are 11, 2×11, $2 \times 2 \times 11, 2 \times 2 \times 2 \times 11$ and $2 \times 2 \times 2 \times 4 \times 11$ i.e., 11, 22, 44, 88 and 352 respectively, with total number of gold coins = $11 + 22 + 44 + 88 + 352 = 517$.

40. Sum = Product

There can be different ways of solving this puzzle. One of the ways is explained below:

Let a, b and c be the cost of items expressed in cents, then

$$a + b + c = 549$$

$$a \times b \times c = 5490000 = 3 \times 3 \times 61 \times 2^4 \times 5^4$$

Since 61 is the largest single factor of abc, the cost of one of the items say a is 61 or it's multiple i.e., 61n, where n = 1 to 8 as a < 549 $\Rightarrow$ a < 9 × 61.

Since 7 is not a factor of abc, so $a \neq 7 \times 61$. The remaining possibilities for the value of a and corresponding values of b + c and b × c is shown in Table 4.14.

Table 4.14 .

n	a	b + c	b × c	Remarks
1	61	$549 - 61 = 488$	$3 \times 3 \times 2^4 \times 5^4$	$(b + c) < 5^4$
2	$61 \times 2 = 122$	$549 - 122 = 427$	$2 \times 3 \times 2^3 \times 5^4$	$(b + c) < 5^4$
3	$61 \times 3 = 183$	$549 - 183 = 366$	$3 \times 2^4 \times 5^4$	$(b + c) < 5^4$
4	$61 \times 4 = 244$	$549 - 244 = 305$	$3 \times 3 \times 2^2 \times 5^4$	$(b + c)$ is a multiple of 5
5	$61 \times 5 = 305$	$549 - 305 = 244$	$3 \times 3 \times 2^4 \times 5^3$	$(b + c) > 5^3$
6	$61 \times 6 = 366$	$549 - 366 = 183$	$3 \times 2^3 \times 5^4$	$(b + c) < 5^4$
8	$61 \times 8 = 488$	$549 - 488 = 61$	$3 \times 3 \times 2 \times 5^4$	$(b + c) < 5^4$

From Table 4.14, it can be seen that for n = 1, 2, 3, 6 and 8, (b + c) is less than 5^4 and 5 is not a factor of (b + c), so one of the numbers i.e., either b or c must contain all factors of 5 i.e., 5^4, which is not possible as $(b + c) < 5^4$.

Let's now consider the remaining possibilities i.e., for n = 4 and n = 5.

For n = 5, b + c = 244 and b × c = $3^2 \times 2^4 \times 5^3$

Since b + c is not a multiple of 5, either b or c must be 5^3 or its integer multiple.

Let b = 5^3k, so c = 244 − 5^3k $\Rightarrow$ k = 1.

So, c = 244 − 125 = 119 = 7 × 17.

Since factors of c i.e., 7 and 17 are other than 2, 3 and 5, this is not possible.

For n = 4, b + c = 305 and b × c = $3^2 \times 2^2 \times 5^4$

Since b + c is a multiple of 5 and not a multiple of 25, either one of the numbers is a multiple of 5 and the other is a multiple of 5^3.

Let b = 5^3m and c = 5n, so b × c = 5^4 × m × n and m × n = $3^2 \times 2^2$ = 36,

b + c = 5 × (5^2 × m + n) = 305 $\Rightarrow$ 25 × m + n − 61 = 0.

So, m × n = m (61 − 25 × m) = 36

$\Rightarrow$ 25 m^2 − 61 m + 36 = 0

$\Rightarrow$ 25 m^2 − 25 m − 36 m + 36 = 0

$\Rightarrow$ 25 m (m − 1) − 36 (m − 1) = 0

$\Rightarrow$ m = 1 (integer value), so n = 36

$\Rightarrow$ b = 5^3 = 125 and c = 5 × 36 = 180

The cost of items in dollars is obtained by dividing the values of a, b and c by 100. So, the cost of the three items is 2.44, 1.25 and 1.80 dollars.

We can see that 2.44 + 1.25 + 1.80 = 5.49 and 2.44 × 1.25 × 1.80 = 5.49.

Comments: One of the following important results can be used in such types of puzzles to eliminate some of the alternatives given in Table 4.14.

For a given sum of two numbers say x and y i.e., x + y, the largest product of these two numbers i.e., x × y = $(\frac{x+y}{2})^2$.

The largest product is achievable when x = y.

We can see from Table 4.14 that for given b + c, the largest possible value of b × c i.e. $(\frac{b+c}{2})^2$ is less than the actual value of b × c given in table for n = 1, 5, 6 and 8. So these values of n can be straightway eliminated.

This puzzle can also be solved by splitting the factors of a × b × c = 2 × 2 2 × 2 × 3 × 3 × 5 × 5 × 5 × 5 × 61 into three sets in such a way that the sum of three sets i.e., a + b + c = 549.

Let us take one more example of a similar puzzle, where the sum and product of three items are 5.55 instead of 5.49.

In this case a + b + c = 555, where a, b and c are cost of item in cents.

a × b × c = 5550000 = 3 × 37 × $2^4 \times 5^5$

It can be noted that 37 is the largest single factor, so the cost of one of the items say a is 37n where n = 1 to 14 as a < 15 × 37 (i.e., 15 × 37 = 555). Since 7, 9, 11, 13 and 14 are not a factor of a × b × c so n ≠ 7, 9, 11, 13 or 14.

For n = 1, 2, 3, 4, 6, 8 and 12, (b + c) < 5^5 and 5 is not a factor of (b + c), so one of the numbers i.e. either b or c must contain all factors of 5 i.e. 5^5 which is not possible as (b + c) < 5^5, so n ≠ 1, 2, 3, 4, 6, 8 or 12. The remaining possible values of n are 5 and 10.

For n = 10, b + c = 185 and b × c = 3 × $2^3 \times 5^4$

It can be seen that for b + c = 185, maximum possible value of b × c as explained above = $(\frac{b+c}{2})^2 = (\frac{185}{2})^2 < 3 \times 2^3 \times 5^4$, so n ≠ 10.

The only remaining value of n is 5, which gives a = 185, b = 250 and c = 120. So, the cost of the three items is 1.85, 2.50 and 1.20 dollars.

It can be noted that by using a combination of strategies as explained above, such types of puzzles can be simplified to a great extent.

41. Investment and Returns

At a first glance, you may be tempted to conclude that the amount received by selling the plot for 12 million rupees be divided between two friends in the ratio of their initial investments i.e. 9:6, so the jeweller shall receive 7.2 million rupees and broker shall receive 4.8 million rupees. But this is not correct. The flaw in the above solution is the fact that one plot kept by each of the two friends has not been taken into consideration. Since the cost of the plot is the same and is not in proportion to their investments, the net investment shall be taken into account to divide the share of receipt.

The cost of three plots = 15 million rupees

Hence cost of each plot = 5 million rupees

So, net investment by jeweller = 9 − 5 = 4 million rupees

Net investment by broker = 6 − 5 = 1 million rupees

Hence the ratio of net investments of jeweller and broker is 4:1. So their share from proceeds of the sale of one plot for 12 million rupees shall be divided in the ratio of 4:1.

Hence jeweller's share = 12 × 4/5 = 9.6 million rupees and broker's share = 12 × 1/5 = 2.4 million rupees.

42. The Grazing Cow's

The important point in solving such a puzzle is to first calculate the quantity of grass growing on one acre per day, as the grass is growing all the time and is to be taken into account.

Let the quantity of grass growing on one acre per day = t times the original quantity of grass when cows start grazing.

In the first field of 30 acres, increase of grass = 30t per day

So, increase of grass in 10 days = 30t × 10 = 300t

Hence total grass grazed by 40 cows (in acres) in first field = original grass + increase in grass = 30 + 300t.

So, the total grass grazed by one cow in the first field in 10 days = $\frac{(30+300t)}{40}$.

Hence grass grazed by one cow in one day in the first field

$$= \frac{(30+300t)}{40 \times 10} \tag{i}$$

Similarly, total grass grazed by one cow in one day in the second field

$$= \frac{42 + 42t \times 12}{50 \times 12} \tag{ii}$$

Since cows eat the same amount of grass every day, the value of t can be obtained by equating Eqs. (i) and (ii)

$$\frac{(30 + 300t)}{40 \times 10} = \frac{42 + 42t \times 12}{50 \times 12}$$
$$\Rightarrow t = 1/18$$

From (i) or (ii), we can compute grass grazed by one cow in one day.

$$= \frac{30 + 300 \times 1/18}{40 \times 10}$$
$$\Rightarrow \frac{3}{40} + \frac{1}{24} = \frac{7}{60}$$

Now from the data given for the third field, we can find the number of cows. Let the number of cows which can eat 63 acres of grass in 15 days be x. So total grass grazed by x cows (in acres) in 15 days is

$$= 63 + 63 \times 15t$$
$$= 63(1 + 15/18)$$
$$= 63 \times 33/18$$

So, grass grazed by one cow in one day, which is the same for all fields as computed earlier is 7/60, so

$$\frac{63 \times 33}{18 \times x \times 15} = \frac{7}{60}$$
$$\Rightarrow x = \frac{63 \times 33 \times 60}{7 \times 18 \times 15}$$
$$\Rightarrow x = 66$$

Hence 66 cows will graze 63 acres of field in 15 days.

43. Labour on a Construction Site

Let there are x men, y women and z children working on the construction site, so

$$x + y + z = 50 \tag{i}$$

$$500\,x + 400\,y + 100\,z = 15000$$

$$\Rightarrow 5x + 4y + z = 150 \tag{ii}$$

From Eqs. (i) and (ii), we get

$$x = 3z - 50 \tag{iii}$$

$$y = 100 - 4z \tag{iv}$$

As per the given condition, a woman is permitted to work only if her husband works, so $y < x$. As per another condition, at least half the men working are accompanied by their wives, so $y > \frac{x}{2}$. This implies that y lies between x and $\frac{x}{2}$. If $y = x$, then from Eqs. (iii) and (iv), we get

$$3z - 50 = 100 - 4z$$
$$\Rightarrow 7z = 150$$
$$\Rightarrow z = 150/7 = 21.4$$

If $y = x/2$, then from Eqs. (iii) and (iv) we get

$$3z - 50 = 2(100 - 4z)$$
$$\Rightarrow 11z = 250$$
$$\Rightarrow z = 250/11 = 22.7$$

So, z lies between 21.4 and 22.7. Since z is an integer, so $z = 22$.
From Eqs. (iii) and (iv) we get $x = 16$ and $y = 12$.
So, 16 men, 12 women and 22 children are working on the day, when the contractor paid 15000 rupees on that day.

44. The Biggest Number Using Three Identical Digits

It can be easily seen that if the digit (say d) is more than 1, then either d^{dd} or d^{d^d} will give the biggest number. Let's find out which of the above two arrangements will give the biggest number for a given digit 'd'.
Let $d^{dd} < d^{d^d} \Rightarrow dd < d^d$
$\Rightarrow (10d + d) < d^d \Rightarrow 11d < d^d \Rightarrow 11 < d^{d-1}$
So, if $d^{d-1} > 11$, then d^{d^d} will give the biggest number using three d's.

$$
\begin{aligned}
\text{For}\quad d &= 1, \quad d^{d-1} = 1 < 11 \\
d &= 2, \quad d^{d-1} = 2 < 11 \\
d &= 3, \quad d^{d-1} = 9 < 11 \\
d &= 4, \quad d^{d-1} = 64 > 11
\end{aligned}
$$

So $d^{d-1} > 11$, if $d > 3$.

Hence for $d = 4$, d^{d^d} i.e. 4^{4^4} will give the biggest number using three 4's. In fact, for $d = 5, 6, 7, 8$ or 9, the arrangement d^{d^d} will give the biggest number.

Comment: It can be interesting to find the biggest number using four identical digits. For example, the biggest number using four 3's is $3^{3^{3^3}}$, whereas the biggest number obtained using four 7's is $7^{7^{7^7}}$.

45. Five Jealous Husband and Their Wives

Let five men be designated as A, B, C, D and E. Their wives are designated as a, b, c, d and e respectively. The details of crossings are tabulated in Table 4.15.

Table 4.15 .

Crossing number	Left shore of river	Crossing	Right shore of river	Remarks
Initially	A B C D E a b c d e	–	–	–
1	A B C D E d e	$\xrightarrow{abc}$	a b c	3 wives cross
2	A B C D E a d e	$\xleftarrow{a}$	b c	One wife return
3	A B C D E e	$\xrightarrow{ad}$	a b c d	2 wives cross
4	A B C D E a e	$\xleftarrow{a}$	b c d	One wife return
5	A E a e	$\xrightarrow{BCD}$	B C D b c d	3 husbands cross whose wives crossed earlier and are at right shore
6	A B E a b e	$\xleftarrow{Bb}$	C D c d	One couple return
7	a b e	$\xrightarrow{ABE}$	A B C D E c d	3 husbands cross
8	a b c e	$\xleftarrow{c}$	A B C D E d	One wife return
9	c	$\xrightarrow{abe}$	A B C D E a b d e	3 wives cross
10	ac	$\xleftarrow{a}$	A B C D E b d e	One wife return
11	–	$\xrightarrow{ac}$	A B C D E a b c d e	Finally, two wives cross

So, eleven crossings are required by which five jealous husbands and their wives can cross the river. There can be other solutions also as shown in the comments section.

Comments: If husbands are not jealous, then only nine crossings are sufficient. In this case, three people cross and one returns in every crossing. So, in 8 crossings 8 persons are transferred and in the last crossing, three persons cross. Other solutions with eleven crossings can be obtained as follows:

(i) In the second crossing, instead of one wife return, it can be changed to "two wives return" and in the third crossing, instead of "2 wives cross" can be changed to "three wives cross", without affecting further results.

(ii) Similar changes in the 10th and 11th crossing will give another solution.

(iii) Another solution with eleven crossings can be obtained by making the following changes in the 10th and 11th crossings.

| 10 | Cc | C | A B C D E a b d e | A husband whose wife is at other shore return |
| 11 | – | Cc | A B C D E a b c d e | Couple crosses |

If x denotes the capacity of the boat, the number of minimum crossings n required for N couples to cross is given in Table 4.16.

Table 4.16 .

N	3	4	5
x	2	3	3
n	11	9	11

It will be interesting to try for $N > 5$, where one can find that no solution can be obtained if the boat capacity is 3 or less.

46. The Look and Say Sequence

Let's first find the logic of the terms in the given sequence.

The first term is 0. Further terms of the sequence are generated by looking at the previous term and describing it as follows:

'0' is described as 'one zero' i.e., '10' which becomes the next term. Now '10' is described as 'one one one zero' i.e., '1110', which becomes the next term. '1110' is described as 'Three one one zero' i.e., '3110' which becomes the next term.

Proceeding in this way, one can obtain the required two terms of the given sequence as follows:

13211321322110

1113122113121113222110

Comments: Such type of sequences called the look and say sequences were introduced by John Conway. The first term of the sequence was taken as 1 and the sequence is

1, 11, 21, 1211, 111221 ...

The interesting properties of such sequences are:

If the first term of the sequence is the digit 'd', where d is any digit from 0 to 9,

(i)   The last digit of every term of the sequence will always be d.

(ii) The sequence is infinite.

(iii) Except for the first term, which is 'd', and the last digit of every term i.e., d, no digit other than 1, 2 and 3 appears in the sequence.

(iv) If the sequence is generated with any digit d from 0 to 9 except 1, as the first term, all sequences are identical except the first term and last digit of every term. The form of these sequences will be d, 1d, 111d, 311d, 13211d, 111312211d, 31131122211d, 1321132132211d, ...

It can be noted that, except d, no digit other than 1, 2 and 3 appears anywhere in the sequence.

(v)   No digit occurs more than three times consecutively in any term of any sequence.

47. Thirty-Six Inch Scale with Minimum Markings

Assuming that start marking i.e., '0' and end marking i.e., the length of scale is known, the minimum number of other markings required to measure all integral lengths from 1 to 12 inches in a 12-inch scale is 4, and three solutions are markings at a, b, c and d (refer Fig. 4.22), which are:

$$1, 2, 3 \text{ and } 8 \text{ inches}$$

$$1, 2, 6 \text{ and } 8 \text{ inches}$$

$$1, 3, 7 \text{ and } 11 \text{ inches}$$

Fig. 4.22 .

Similarly, for a 36-inch scale, 8 markings are required to measure any integral length from 1 to 36 inches. These eight markings divide the 36 inches distance into the following nine sections i.e., 1, 2, 3, 7, 7, 7, 4, 4 and 1.

The 8 markings are as shown in Fig. 4.23.

Fig. 4.23 .

These are at 1, 3, 6, 13, 20, 27, 31 and 35 inches from one end.

Another solution to this puzzle can be obtained by translation and the eight markings, in that case, are at 1, 5, 9, 16, 23, 30, 33 and 35 inches from one end.

Comments: I have tried to find the minimum values of markings required that can measure all integral lengths of scale for a certain range. For example, five markings (excluding start and end markings of the scale) are sufficient to measure any integral length from 1 to 14 inches, 1 to 15 inches, 1 to 16 inches and 1 to 17 inches. That means for a scale of 14 inches to 17 inches, only 5 markings are required to measure any integral length from one inch to the length of the scale. Table 4.17 gives the minimum markings required (excluding start and end markings of the scale) for a certain range of scale.

Table 4.17 .

Number of markings	Range of scales (length in inches)
1	2–3
2	4–6
3	7–9
4	10–13
5	14–17
6	18–23
7	24–29
8	30–36
9	37–43
10	44–50
11	50–58

So, for a 43-inch scale, nine markings (excluding start and end markings) are sufficient to measure all integral lengths from 1 to 43 inches and markings are at 1, 3, 6, 13, 20, 27, 34, 38 and 42 inches.

The other solution can be obtained by translation and is given by 1, 5, 9, 16, 23, 30, 37, 40 and 42 inches.

It can be an interesting pastime to find the number of possible different solutions for different ranges of scales for minimum markings. The possibility to search for the higher ranges of scales by observing the patterns/symmetries in the solution of lower ranges can also be explored.

48. Number of Triangles

For the maximum number of triangles that can be formed from any given number of straight lines, no two lines must be parallel and no three lines shall intersect at a common point. Considering the above conditions, four triangles

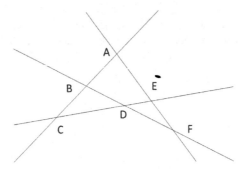

Fig. 4.24 .

can be formed with four straight lines as shown in Fig. 4.24.

The four triangles formed are BCD, EDF, ACE and ABF. If there are N straight lines, such that no two lines are parallel and no three or more lines intersect at a common point, the maximum number of triangles that can be formed is $^{N}C_3$, which is the number of ways of choosing three lines out of N lines, to be the sides of the triangles.

$$^{N}C_3 = \frac{N!}{3!(N-3)!}$$

If N = 4, No. of triangles = $^{4}C_3 = \frac{4\times3\times2\times1}{3\times2\times1\times(4-3)!} = 4$.

Similarly, for N = 5, 6 and 7, the number of triangles formed is 10, 20 and 35. So, 35 triangles can be formed with seven straight lines.

49. The Ants on a Stick

(a) Let's consider four ants A, B, C and D walking in the directions as shown in Fig. 4.25.

Note that the speed of all ants is the same.

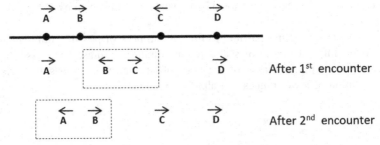

Fig. 4.25 .

The first encounter occurs between B and C. The position of ants after the first encounter is shown in Fig. 4.25. The second encounter occurs between A and B. The position of ants after the second encounter is shown in Fig. 4.25. There will be no further encounters so a total of two encounters take place.

Now let's consider all seven ants as given in the puzzle. The position of ants after each stage of the encounter is shown in Fig. 4.26. As can be seen from Fig. 4.26, three encounters i.e., between B and C, D and E, F and G take place in the first stage.

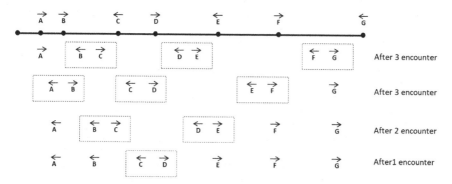

Fig. 4.26 .

Similarly, three encounters i.e., between A and B, C and D, E and F take place in the second stage. Two encounters i.e., between B and C, D and E in the third stage and finally one encounter between C and D in the final stage takes place resulting in a total of nine encounters.

Let's now see, how to determine the number of encounters, if there are many ants on the stick.

There are two ways:

(i) Let's consider right-moving ants i.e. A, B, D and F

The number of encounters = The number of left-moving ants right of A
+ The number of left-moving ants right of B
+ The number of left-moving ants right of D
+ The number of left moving ants right of F
= 3 + 3 + 2 + 1 = 9

The sum is to be taken for each of the right moving ants.

The same result will be obtained if we consider left-moving ants i.e., C, E and G.

The number of encounters = The number of right-moving ants left of C
+ The number of right-moving ants left of E
+ The number of right-moving ants left of G.
= 2 + 3 + 4 = 9.

(ii)   The number of encounters can also be obtained by the graphical method as
follows:

Draw all ants on the horizontal axis. Draw lines at 45° from the position of each
ant in the direction of their initial motion like AA, BB ... etc. as shown in
Fig. 4.27.

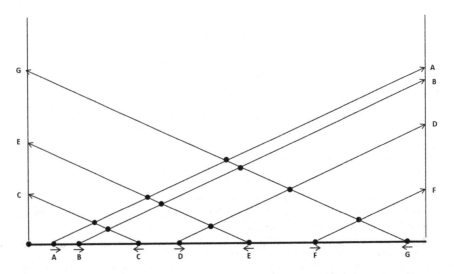

Fig. 4.27 .

Count the number of intersection points as shown in Fig. 4.27. These are nine
in number which gives the number of encounters. Both the methods given
above can be used if there are many ants on the stick.

(b)   This will depend upon the initial position of the first and last ant.

Consider two ants A and G moving in the opposite directions on a stick of
length 1 units as shown in Fig. 4.28.

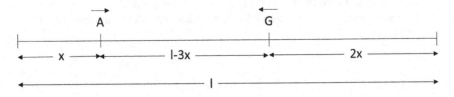

Fig. 4.28 .

Let the initial position of ant A (moving right) is at a distance of x units from one end and ant G (moving left) is at a distance of 2x units from the other end. In this case, the maximum distance covered by the ant (i.e., G) is (1 − 3x) + 2x = 1 − x, which is not equal to the length of the stick but is the largest initial distance of any ant (i.e., A) from far end of the stick.

The time required by the last ant to fall off the stick will be governed by the maximum initial distance of any ant from the far end of the stick. So, find the ant which is at the farthest point from the end it is facing.

In our case, the maximum initial distance of any ant from the far end of the stick is 100 − 5 = 95 cm.

So, the time required for the last ant to fall off = 95/1 = 95 seconds. Note that the ant which is at a maximum initial distance from the far end (i.e., ant A) is not the last ant to fall off the stick. It is only governing the time required by the last ant to fall off the stick. The above analysis holds for any number of ants on the stick.

(c) Now let's identify the ant that falls off the stick in the last.

The key point to note is that the order of ants doesn't change as they pass each other.

(i)   At the start, four ants are walking to the right, so four ants will fall off the stick on the right end. Since ants cannot pass any other ant, four right ants will fall off to the right end. So, the ant that is positioned fourth to the right will fall last i.e., ant D.

(ii) Similarly, three ants are walking to the left initially, so three ants will fall off the stick on the left end. Since ants cannot pass any other ant, so three left most ants will fall off to the left end. Hence the ant that is positioned third to the left will fall last i.e., ant C.

Now out of (i) and (ii), we have to find which ant i.e., C or D will fall in the last. If the left most ant facing right (i.e., ant A) is closer to the edge (the distance of A from the near end is 5 cm) than the right most ant facing left (i.e., ant G with distance from its near end as 20 cm), the last ant will fall off to the right end. If the last ant falls off to the right end, then it is ant D else it is ant C.

If the edge distance is the same in both cases, then both ants i.e., C and D will fall off simultaneously to the left and right respectively.

Since in the given puzzle, the edge distance of ant A is less than the edge distance of ant G, so the last ant will fall to the right and it is ant D.

50. **Winner of the Year**

The first team i.e., team A can ensure its victory by starting with the date 20 January and successively adhering to the following winning dates to ensure its victory.

21 February, 22 March, 23 April, 24 May, 25 June, 26 July, 27 August, 28 September, 29 October and 30 November.

As a strategy winning team must start with the date of 20 January and then irrespective of what the opponent team selects, the winning team must select one of the dates given above at its turn to ensure its victory. If the winning team restricts to the details as above at his turn, the opponent team will not be able to select any of the above dates.

To understand, how these winning dates have been arrived at, start from the final winning date i.e., 31 December and proceed backward.

The winning team (say team A) selects 30 November on his pre-winning turn to force the opponent team (say team B) to select the only option available i.e., 30 December, which finally gives chance to team A to select a winning date of 31 December.

Let team A select 29 October, in this case, the available option with the opponent team is 30 October, 31 October, 29 November or 29 December.

If the opponent team selects 31 October or 29 December, team A can directly select the winning date i.e., 31 December. If the opponent team selects 30 October or 29 November, team A can again select one of the winning dates i.e., 30 November, which ensures the victory of team A.

Proceeding similarly, one can find that if team A adheres to the winning dates mentioned above, team A can force the opponent team to select dates other than the winning dates mentioned above and ensure its victory.

51. Careless Bank Cashier

From the given data, the important observations are:

(i) Since 1 rupee = 100 paisa, so

$$y < 100 \text{ and } x < 100$$

(ii) y > x, otherwise the cashier would have returned a lesser amount.

(iii) Since the cashier has returned an amount equal to more than twice the original amount of the cheque, so y > 2x.

The original amount of the cheque = x rupees and y paisa.
The incorrect amount received = y rupees and x paisa.
After giving 20 paisa to a beggar, the balance amount left is double the amount of the original cheque, so

$$100y + x - 20 = 2(100x + y)$$

$$\Rightarrow 98y = 199x + 20$$

$$\Rightarrow y = 2x + \frac{3x + 20}{98} \text{ (This verifies } y > 2x)$$

Also, since y < 100, so x < 50.
As x and y are integer values, so 3x + 20 must be a multiple of 98.
Let 3x + 20 = 98k

$$\Rightarrow x = \frac{98k - 20}{3}$$

Since x < 50, so k < 2

For k = 0, x is negative.

For k = 1, x = 26 and y = 2 × 26 + $\frac{3 \times 26 + 20}{98}$ = 53.

So, the original amount of the cheque is 26 rupees and 53 paisa.

52. Five Working Days a Week

There are 31 days in March and 30 days in April in any year.

There can be a maximum of 5 Saturdays and 5 Sundays in March and a minimum of 4 Saturdays and 4 Sundays in March. So, the minimum and maximum working days in March can be 21 and 23 respectively. Similarly, working days in April can be 20 to 22. If the number of working days in March and April are equal, then it can be either 21 or 22.

For 21 working days in March, there must be 5 Saturdays and 5 Sundays in March. For this to happen, either 1st March shall be Friday or Saturday.

If 1st March is Friday, then 1st April will fall on Monday but this will result in 22 working days in April as against 21 working days in March.

If 1st March is Saturday, then 1st April will fall on Tuesday and this will also result in 22 working days in April as against 21 working days in March. So, for equal working days in March and April, 21 working days are ruled out.

For 22 working days in March, either Saturday or Sunday (but not both) must occur five times in March, for this to happen, 1st March shall fall on either Sunday or Thursday. If 1st March is Sunday, then 1st April will fall on Wednesday and this will give 22 working days in April also, which is equal to the number of working days in March.

If 1st March is Thursday, then 1st April will fall on Sunday resulting in 21 working days in April as against 22 working days in March.

Hence for equal working days in March and April, there will be 22 working days in March and April and the first April will fall on Wednesday.

53. Defective Coin Among 24 Coins

Divide 24 coins into groups of 8 coins each say A, B and C. Balance two groups of 8 coins each say A and B.

If A > B, this indicates B is lighter hence group B contains the defective coin.

If A < B, group A contains the defective coin.

If A = B, group C contains the defective coin.

So, in one balancing, the problem reduces to one-third of the coins i.e., from 24 to 8 coins only.

In the second balancing, take the group of 8 coins containing the defective coin. Divide this group into three groups say X, Y and Z containing 3, 3 and 2 coins respectively. Balance two groups of 3 coins each i.e., X and Y.

If X > Y, this indicates Y is lighter hence group Y contains the defective coin.

If X < Y, group X contains the defective coin.

If X = Y, group Z contains the defective coin.

In the third balancing, pick any two coins say a and b from the group containing the defective coin and balance these two coins.

If a < b, a is the defective coin.

If a > b, b is the defective coin.

If a = b, the remaining third coin is defective.

If X = Y in the second balancing, then only 2 coins are left in group Z, which can be balanced to identify the defective one.

So, three balancings are sufficient to identify the defective coin among 24 coins. The same methodology is applied if the given defective coin is heavier than the other coins.

Comments: It can be seen that even if 27 coins were given including one defective coin (either lighter or heavier which is known), the solution can be obtained in three balancings only. In every balancing three groups of an equal number of coins are taken.

In general n balancings are sufficient to identify the defective coin from $N = 3^n$ number of identical coins containing one defective coin (whether lighter or heavier which is known).

The same number of balancings are required for N coins if $3^{n-1} < N \leq 3^n$. So, the minimum number of balancings to identify the defective coin from a group of N coins is equal to $[\log_3 (N - \frac{1}{2})] + 1$, where the square bracket denotes the integer part of the number in square brackets.

54. Two Trains and a Bird

One way to solve this problem is to calculate the distance travelled by the bird in each cycle before reversing its path and then summing the distances together which is an infinite series. However, this is not a good approach.

The simple approach to solve this puzzle is to find the total time for which the bird has travelled before crashing. Let this time is T.

Since the velocity of a bird say V is known, one can find the total distance travelled by a bird, say D in time T i.e.

$$D = VT$$

Time T can easily be calculated, as it is the total time taken by two trains in travelling before the crash.

Let the velocity of train A is V_A and train B be V_B.

Let the initial distance between two trains is d.

Distance travelled by train A in time t = t × V_A

Distance travelled by train B in time t = t × V_B

Total distance travelled by both trains in time t is

$$d = t \times V_A + t \times V_B \Rightarrow t = \frac{d}{V_A + V_B}$$

$$\Rightarrow t = \frac{500}{100 + 150} = 2 \text{ hours}$$

Since T = t, so D = 200 × 2 = 400 km.
So, the total distance travelled by bird before the crash is 400 km.

Comments: Since both trains are travelling in the opposite direction towards each other, their relative velocity is 100 + 150 = 250 km/h. So, time taken by trains before the crash = 500/250 = 2 hours, which is also the time taken by the bird in travelling before the crash. So, the distance covered by the bird before the crash = 200 × 2 = 400 km.

55. Basket and Eggs

Let there are N eggs in the basket, so N must be divisible by 7, say N = 7 m. Since on dividing N by 2, 3, 4, 5 and 6, a remainder one less than the divisor is obtained, so N + 1 must be divisible by 2, 3, 4, 5 and 6.
LCM of 2, 3, 4, 5 and 6 = 60.
Hence any multiple of 60, say 60K minus 1, will satisfy the conditions except the condition of divisibility by 7. So, to satisfy all conditions, the following two equations must be satisfied i.e.
N = 7m and N = 60K − 1
$\Rightarrow m = \frac{60K-1}{7} = 8K + \frac{4K-1}{7}$
The smallest positive integer K, such that m is also positive integer is K = 2, which gives
$m = 8 \times 2 + \frac{4 \times 2-1}{7} = 17$. Hence N = 7 × 17 = 119.
So, 119 is the smallest number of eggs that must have been in the basket.
The general solution will be N = 119 + 60 × 7n = 119 + 420n, where n = 0 gives the smallest solution.

Comments: Another variant of the puzzle is: "If the eggs in the basket are taken out 2, 3, 4, 5 and 6 at a time, there remains one egg every time. However if 7 eggs are taken out, none are left in the basket". This also can be solved similarly as follows:
LCM of 2, 3, 4, 5 and 6 = 60
Now instead of one less, here remainder is one more than the divisor, so N − 1 must be a multiple of 60. Hence N = 60K + 1, N = 7m
$\Rightarrow m = 8K + \frac{4K+1}{7}$
The smallest positive integer value of K, such that m is also a positive integer is K = 5, which gives m = 43. So, N = 7 × 43 = 301 is the smallest number of eggs that could be in the basket. The general solution in this case is N = 301 + 60 × 7n, where n = 0, 1, 2 …
So, N = 301, 721, 1141 … etc.

56. Ranking Large Numbers

Before attempting the puzzle, it is important to know the following well-known result.

As n tends to infinity, the value of $(1 + \frac{1}{n})^n = e = 2.718 \ldots$

For $n = 1$, $(1 + \frac{1}{n})^n = 2$

For $n = \infty$, $(1 + \frac{1}{n})^n = e$

So, $2 \le (1 + \frac{1}{n})^n < 3$

Now let's compare 1000^{1000} and 1001^{999} by converting these in the form of $(1 + \frac{1}{n})^n$ as follows:

$$\frac{1001^{999}}{1000^{1000}} = \left(\frac{1001}{1000}\right)^{1000} \times \frac{1}{1001}$$

$$= \frac{1}{1001} \times \left(1 + \frac{1}{1000}\right)^{1000} < \frac{3}{1001} < 1$$

Hence $1001^{999} < 1000^{1000}$

Let's now compare 1000^{1000} and 999^{1001}

$$\frac{1000^{1000}}{999^{1001}} = \frac{1}{999} \times \left(1 + \frac{1}{999}\right)^{999} \times \frac{1000}{999} < \frac{1000 \times 3}{999 \times 999} < 1$$

So, $1000^{1000} < 999^{1001}$

Hence 999^{1001} is the greatest number and 1001^{999} is the least number among the given three numbers.

57. A Unique Polydivisible Number

The following divisibility rules must be kept in mind.

(i)   If the last n digits are divisible by 2^n, then the whole number is divisible by 2^n, so if the last two digits of a number are divisible by 4 (i.e., 2^2) then the whole number is divisible by 4. Similarly, if the last 3 digits of a number are divisible by 8 (i.e., 2^3), the whole number is divisible by 8 and so on.

(ii) If the sum of digits of a number is divisible by 3, then the number is divisible by 3 and vice versa.

(iii) If the sum of digits of a number is divisible by 9, then the number is divisible by 9 and vice versa.

Let the ten-digit number be ABCDEFGHIJ.

(i)   Since the ten-digit number is divisible by 10, so J = 0.
      Similarly, the first 5-digit number i.e., ABCDE is divisible by 5, so E is either 0 or 5 but each digit can only be used once and J = 0, so E = 5.

(ii) First 2-digit, 4-digit, 6-digit and 8-digit numbers are divisible by 2, 4, 6 and 8 respectively, so B, D, F and H are even digit except 0.

(iii) Remaining digits i.e., 1, 3, 7 and 9 must occupy the remaining places, so A, C, G and I must be odd except 5.

(iv) Since ABCD is divisible by 4, the CD must be divisible by 4. Since C can be 1, 3, 7 or 9 and D can be 2, 4, 6 and 8, for CD to be divisible by 4, possible values of CD can be 12, 16, 32, 36, 72, 76, 92 or 96 only. Hence D can only be 2 or 6.

(v) ABC is divisible by 3, so A + B + C must be divisible by 3.
ABCDEF is divisible by 6 and ABC is divisible by 3, so DEF shall be divisible by 3, which implies that D + E + F must also be divisible by 3.
Since E = 5, D = 2 or 6, F = 2, 4, 6 or 8 and D + E + F is divisible by 3, so possible values of DEF are 258 or 654 only, as each digit can appear only once. So, F can be 4 or 8 only.

(vi) ABCDEFGH is divisible by 8, so FGH must be divisible by 8.
Now F = 4 or 8, G = 1, 3, 7 or 9 and H = 2, 4, 6 or 8.
So, the possible values of FGH which is divisible by 8 are 416, 432, 472, 496, 816, 832, 872 or 896.
This indicates that the value of H can be 2 or 6.

(vii) Consider even positions, where we have
J = 0, H = 2 or 6, F = 4 or 8, D = 2 or 6, B = 2, 4, 6 or 8
Since both H and D are 2 or 6, so no other number can have 2 or 6, hence B = 4 or 8.

(viii) Let's now find possibilities for ABC.
A = 1, 3, 7 or 9

B = 4 or 8

C = 1, 3, 7 or 9.

Since ABC is divisible by 3 and no digits are repeated, so possible values of ABC are 147, 183, 189, 381, 387, 741, 783, 789, 981 and 987.

(ix) ABCDEFG is divisible by 7. Considering possible values of ABC, DEF, and FGH computed above, the possible values of ABCDEFGH where no digit is repeated are,

14725896

18365472

18965432

18965472

38165472

74125896

78965432

98165432

98165472

98765432

Out of the 10 values given above, 38165472 is the only number, where the number formed from the first seven digits is divisible by 7. So ABCDEFGH = 38165472.

(x) The only remaining digit 9 takes the place of the only remaining unknown letter i.e. I, So I = 9. So, the ten-digit number is 3816547290.

Comments: A number that includes all ten digits from 0 to 9 is called a pandigital number. A number is called polydivisible, if the number formed from its first n digits is divisible by n. For a given base, there are only a finite number of polydivisible numbers. In base-10, the largest polydivisible number is a 25-digit number i.e.

$$3608528850368400786036725$$

There are 2492 polydivisible numbers with 10 digits but only one 10-digit polydivisible number is pandigital so it is unique.

The puzzle can be modified to find nine-digit polydivisible numbers containing digits from 1 to 9 exactly once. The obvious answer is 381654729 which is obtained just by deleting 0 as the last digit of the 10-digit unique polydivisible number as computed above.

58. Handless Clock Chimer

When a clock strikes 3 o'clock, it strikes 3 times and takes 3 seconds from 1st strike to 3rd strike. So, the time interval between two strikes is 3 seconds/2 = 1.5 seconds, which is constant between any two strikes.

At 10 o'clock, the clock strikes 10 times, which has nine intervals of time, so will take $9 \times 1.5 = 13.5$ seconds. So, after the first strike, it takes 13.5 seconds to strike at 10 o'clock but at this point, one is not sure whether it is 10 o'clock or not and whether further strikes will continue or not. So, one has to wait for one more time interval of 1.5 seconds and if there is no further strike, then one can be certain that the time is 10 o'clock.

So, the total time required after hearing the first strike, if it is certainly 10 o'clock is

$$13.5 \text{ sec.} + 1.5 \text{ sec.} = 15 \text{ seconds}$$

Comments: If the time taken in striking which was neglected in the puzzle is also required to be considered then proceed as follows.

Let it takes s seconds in striking once and t seconds is the time interval between two strikes.

The clock takes 3 seconds to strike 3 o'clock, so it takes time in 3 strikes and two-time intervals between 3 strikes.

$$\text{So}, 3s + 2t = 3 \tag{i}$$

For 10 o'clock, there are 10 strikes and nine-time intervals, so total time taken in striking 10 o'clock is say x, then

$$10s + 9t = x \qquad\qquad\text{(ii)}$$

From (i) t = 3 (1 − s)/2
Substituting value of t in Eq. (ii), we get

$$x = 10s + 9 \times 3(1 - s)/2$$
$$= 13.5 - 3.5s$$

So, if s = 0, x = 13.5 seconds,
If s is given, one can find the total time taken i.e., x in striking 10 o'clock.

59. Mango Sharing

Let the minimum number of mangoes ordered = a
After discarding one rotten mango, share taken by person A = (a − 1)/3
Let the left-over mangoes are b, so b = (a − 1) − (a − 1)/3

$$\Rightarrow a = \frac{3b}{2} + 1$$

Similarly, after discarding 2 mangoes, share taken by person B = (b − 2)/3, let
the left-over mangoes be c, so $c = (b - 2) - \frac{b-2}{3}$
$\Rightarrow b = \frac{3c}{2} + 2$
After discarding 3 mangoes, share taken by person C $= \frac{c-3}{3}$
Let the leftover mangoes now are d, so $d = (c - 3) - \frac{c-3}{3}$

$$\Rightarrow c = \frac{3d}{2} + 3$$

For number of mangoes a, b, c and d to be an integer value, d, c and b must be
even.
For a to be minimum, d shall be minimum. Let minimum even value of d = 2x, so
$c = \frac{3(2x)}{2} + 3 = 3 (x + 1)$, and x can take any positive integer value i.e., 1, 2, 3,
4 ...
$b = \frac{3(3x+3)}{2} + 2 = \frac{9(x+1)}{2} + 2$, so for b to be an integer value, x must be odd i.e.,
1, 3, 5, 7 ...
$a = \frac{3}{2}[\frac{9(x+1)}{2} + 2] + 1$
$= \frac{27}{4} (x + 1) + 4$
For a to be an integer value, x must be 3, 7, 11, 15 ... i.e., of the form 4 n + 3.
The minimum value of x, which is +ve integer value, odd number and also the
number of the form 4 n + 3, as required by the above condition is satisfied by
x = 3.
For x = 3, d = 6, c = 3 × 3 + 3 = 12, b = $\frac{9}{2}$ (3 + 1) + 2 = 20 and
$a = \frac{3b}{2} + 1 = 31$.

Hence 6 mangoes are left over in the last (as d = 6) and the minimum number of mangoes ordered = 31. The number of mangoes kept by person A, B and C can also be calculated as follows:

No. of mangoes kept by person A = $\frac{(31-1)}{3}$ + 6 = 16, by person B = $\frac{20-2}{3}$ = 6 and by person C = $\frac{(12-3)}{3}$ = 3.

Accounting   6   discarded   mangoes,   total   mangoes   ordered   are 6 + 16 + 6 + 3 = 31.

60. Five Sundays in February

For five Sundays (or any weekday) in February of a year, the year must be a leap year to have 29 days in February. In such cases of five Sundays in February of a year, 1st February must fall on Sunday.

If 1st February falls on Sunday of a year, then 1st January will fall on Thursday in that year.

As per the Gregorian calendar, a leap year occurs when the year is divisible by 4. However, if the year is divisible by 100, it is not a leap year except when it is divisible by 400. So, 2100, 2200 and 2300 are non-leap years, whereas 2000 and 2400 are leap years. 1st February 2020, which is a leap year, falls on Saturday (Given), so 1st January 2020 will fall on Wednesday. So, we have to find the years, when 1st January falls on Thursday.

Every four years, we add 5 days i.e., one day for each year plus one day for the leap year. However, we need to add $7n + 1$ day from 1st January 2020 (Wednesday) to get 1st January on Thursday. For this $5k - 1$ must be a multiple of 7, which gives k = 3. Hence $4 \times 3 = 12$ years from 2020 i.e., in 2032, 1st January will fall on Thursday. Consequently, February in 2032 will contain, five Sundays. Next such leap years can be obtained by adding 28 years. So, 2060 and 2088 will be years when February will contain five Sundays. Since 2100 is not a leap year but falls between 2088 and 2088 + 28 = 2116, the year 2116 will not be the year, where February contains five Sundays. It can be calculated that instead of 2116, the next year to contain 5 Sundays in February will be 2128.

Comments: There are 97 leap years in a 400-year cycle in the Gregorian calendar currently in use. The following years will have five Sundays in February in 400 years cycle.

2004, 2032, 2060, 2088, 2128, 2156, 2184, 2224, 2252, 2280, 2320, 2348 and 2376.

A total of 13 leap years will contain five Sundays every four hundred years. Between century years, there is a gap of 28 years but across a century year, there is a gap of 40 years.

61. **Employment During Covid-19 Crisis**

Let the strength of employees in the company is N, where N < 1000
Since 1/8th, 1/5th, 1/3rd and 1/7th of the employees who got grades A, B, C
and D respectively, are positive integers, so N must be a multiple of 8, 5, 3 and
7. LCM of 8, 5, 3 and 7 = 840, so N = 840.
Number of employees with grade A = $\frac{840}{8}$ = 105,
Number of employees with grade B = $\frac{840}{5}$ = 168,
Number of employees with grade C = $\frac{840}{3}$ = 280 and
Number of employees with grade D = $\frac{840}{7}$ = 120.
So total number of employees with grade A, B, C and D = 105 + 168 + 280 +
120 = 673.
Hence remaining employees who got grade E = 840 − 673 = 167.

62. **Camel and Banana Transportation**

Let A be the starting point i.e., farm and B is the destination i.e., market as
shown in Fig. 4.29.

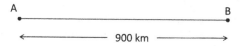

Fig. 4.29 .

Since camel can carry a maximum of 900 bananas at a time and eats one banana
in every km it travels in loaded conditions, if it directly goes from A to B, no
banana will be left at B. So, the bananas will have to be dropped enroute and
stored at some places.
Obviously, at the last store point, the stock shall be 900 bananas so that the
camel can be utilized to its full capacity and the maximum number of bananas
can be transported to market i.e., point B.
Since there are 2700 bananas and the capacity of the camel is 900 bananas, so
three trips will be required by the camel to shift bananas to the first store point.
The first store point shall be so located that from first store point X to second
store point Y, only two trips are required which means from A to X, one trip
banana's equivalent to 900 can be consumed.

(a) Since the camel needs to eat one banana per km only in loaded condition
i.e., forward direction, so camel needs three bananas per km for three
forward trips and no banana will be consumed by the camel in return trips
i.e., empty condition as shown by the dotted line in Fig. 4.30.

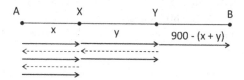

Fig. 4.30

Let distance AX = x km and distance XY = y km. For shifting 2700 bananas from A to X, the camel needs to make three forward trips in loaded condition from A to X and two return trips in empty condition from X to A as shown in Fig. 4.30. In three forward trips, 3x bananas will be consumed by Camel. Also remaining bananas at store point X must be 1800 so that one forward trip is saved between X and Y.

Hence 2700 − 3x = 1800 ⇒ x = 300 km.

To cover the distance XY, only two forward trips and one return trip is needed, so two bananas per km will be consumed by the camel in XY. The available bananas at store point X = 1800 and at Y, it shall be 900 as explained earlier.

So, 1800 − 2y = 900 ⇒ y = 450 km.

Remaining distance YB = 900 − (300 + 450) = 150 km.

In the last segment YB, only one forward trip is required. So, one banana per km will be consumed by the camel. So, bananas remaining at destination point B will be 900 − 1 × 150 = 750.

So, a maximum of 750 bananas can be transported to the market.

(b) In this case, the camel will eat one banana per km in the forward direction i.e., loaded condition, as well as while returning i.e., empty condition. So instead of three bananas per km in earlier case a), five bananas per km will be consumed by the camel in the first segment AX (i.e., three bananas in the forward direction and two bananas while returning) as shown in Fig. 4.31.

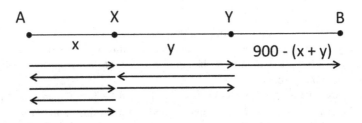

Fig. 4.31 .

So, 2700 − 5x = 1800 ⇒ x = 180 km.

From X to Y, the camel will consume two bananas per km in the forward direction and one banana per km while returning. To keep the remaining bananas at store point X equal to 1800 and at Y equal to 900 as explained earlier in case a),

We have 1800 − 3y = 900 ⇒ y = 300 km.

The remaining distance YB = 900 − (180 + 300) = 420 km.

To cover a 420 km distance, the camel will eat one banana per km in the only trip in the forward direction. So, bananas remaining at destination B i.e., market will be 900 − 1 × 420 = 480 bananas.

So, the maximum number of bananas that can be transported to market, in this case, is 480.

Comments: In such puzzles, decide the intermediate store points in such a way that at the cost of bananas equivalent to the capacity of a camel, all remaining bananas can be shifted in every segment. For example, in the given puzzle, the capacity of the camel is 900 bananas, so in every segment, a camel can consume 900 bananas and transport/shift the remaining ones. Since the number of bananas is three times (i.e., 2700) than the capacity of a camel i.e., 900, a total of three segments i.e., two store points (X and Y) will be required.

For case a), consumption by camel is 3 banana per km (i.e., only in forward direction i.e., loaded condition) in first segment, so

$$3x = 900 \Rightarrow x = 300 \text{ km}$$

Similarly in second segment (only two trips required), so

$$2y = 900 \Rightarrow y = 450 \text{ km}$$

Last segment = 900 − (300 + 450) = 150 km.

For case b), consumption by camel is 5 bananas per km (i.e., 3 bananas for forward direction and 2 bananas for return) in first segment, so

$$5x = 900 \Rightarrow x = 180 \text{ km}$$

Similarly in the second segment, where two trips are sufficient, consumption by camel is three bananas per km (2 bananas for the forward direction and 1 banana for return), so

$$3y = 900 \Rightarrow y = 300 \text{ km}$$

Last segment = 900 − (180 + 300) = 420 km.

63. Distance Between Two Colonies

For a solution, one would attempt to plot the colonies A, B, C, D and E along with distances as shown in Fig. 4.32 and then try to compute the distance AD.

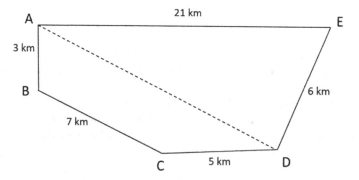

Fig. 4.32 .

A closure look will reveal that from triangle AED, side AD will be more than 15 km, as no side can be more than the sum of the other two sides.

While considering the quadrilateral ABCD, the side AD will be less than 15 km, as AD cannot be more than the sum of the other three sides i.e., 3 + 7 + 5 = 15. So, this contradiction will force you to think again.

It can be seen that, one of the sides i.e., AE = 21 km, is the sum of remaining four sides i.e., AB + BC + BD + DE = 3 + 7 + 5 + 6 = 21 km.

Hence the five colonies A, B, C, D and E will lie on a straight line as shown in Fig. 4.33 and not on vertices of a pentagon as shown in Fig. 4.32.

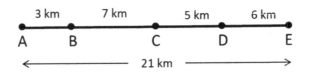

Fig. 4.33 .

So, the distance between colony A to D = AB + BC + CD = 3 + 7 + 5 =15 km.

64. Bachet's Weights

(a) In this case, weights can be placed in one scale pan only and in another scale pan, only the object to be weighed is placed.

Let's start with two smallest weights i.e., 1 kg and 2 kg. With these two weights, we can weigh objects of 1 kg, 2 kg and 1 + 2 = 3 kg. For weighing objects of higher weights, let's add next weight i.e., 4 kg. So now, one can weigh 4 kg, 4 + 1 = 5 kg, 4 + 2 = 6 kg, and 4 + 3 = 7 kg. Hence with three weights i.e., 1 kg, 2 kg and 4 kg, one can weigh any object weighing up to 7 kg.

Proceeding similarly, the next weight required to be added is the previous weight (i.e., 7 kg) + 1 kg = 8 kg and with this, we can weigh up to twice this weight − 1, so with the addition of 8 kg weight, we can weigh up to 2 × 8 − 1 = 15 kg. The next two weights can be added accordingly i.e., 16 kg and 32 kg.

With these 6 weights i.e., 1, 2, 4, 8, 16 and 32 kg, we can weigh all integral weights up to 2 × 32 − 1 = 63 kg. For weighing 64 kg or higher, we need to add one extra weight. If we add 64 kg, we can weigh up to 63 + 64 = 127 kg. But since we need to weigh up to 100 kg only, the addition of a weight of 100 − 63 = 37 kg is sufficient to weigh all weight up to 100 kg. Hence the addition of any weight from 37 kg to 64 kg will be sufficient to weigh all weights up to 100 kg.

So, the required set of weights to weigh any integral weight up to 100 kg (when weights can be placed in one pan only) is 1 kg, 2 kg, 4 kg, 8 kg, 16 kg, 32 kg and any one weight from 37 kg to 64 kg.

(b) In this case, more weighing can be done with the given number of weights, as weights can also be put on a scale pan containing the objects to be weighed.

Starting with two weights i.e., 1 kg and 3 kg, we can weigh 1 kg, 2 kg (i.e., $3 - 1 = 2$), 3 kg and 4 kg (i.e., $1 + 3 = 4$).

For weighing next weight i.e., 5 kg, we need to add another weight. Since weights can be placed on both pans, next weight selected is not 5 kg but 4 kg + 5 kg = 9 kg. So, with three weights i.e., 1 kg, 3 kg and 9 kg, we can weigh up to 13 kg (i.e., 4 kg + 9 kg) as follows:

$5 = 9 - 3 - 1$, $6 = 9 - 3$,

$7 = 9 - 3 + 1$, $8 = 9 - 1$,

$10 = 9 + 1$, $11 = 9 + 3 - 1$,

$12 = 9 + 3$, $13 = 9 + 3 + 1$.

Negative operation in the above relations indicates that this weight is to be placed on a scale pan containing an object. For example, for weighing 11 kg, we have to put 9 kg and 3 kg on one scale pan and 1 kg on another scale pan containing the object.

Next weight to be weighed is 14 kg, for which we have to add 13 kg + 14 kg = 27 kg and with addition of 27 kg weight, we can weigh up to 13 kg + 27 kg = 40 kg.

For example, for weighing 14 kg, 19 kg, 23 kg, 29 kg, 35 kg and 40 kg, we can use the following relations:

$14 = 27 - 9 - 3 - 1$, $19 = 27 + 1 - 9$,

$23 = 27 - 3 - 1$, $29 = 27 + 3 - 1$,

$35 = 27 + 9 - 1$ and $40 = 27 + 9 + 3 + 1$.

So, we can weigh up to 40 kg with the set of four weights i.e., 1 kg, 3 kg, 9 kg and 27 kg.

For weighing 41 kg, we can add 40 kg + 41 kg = 81 kg as discussed earlier and then we can weigh up to 81 kg + 40 kg = 121 kg. Since we need to weigh only up to 100 kg, instead of 81 kg, it is sufficient to add any weight from $100 - 40 = 60$ kg to 81 kg.

For example, for weighing 41 kg, 67 kg, 75 kg and 100 kg, we can use the following relations:

$41 = 60 - 27 + 9 - 1$, $67 = 60 + 9 - 3 + 1$,

$75 = 60 + 27 - 9 - 3$ and $100 = 60 + 27 + 9 + 3 + 1$.

So, the required set of weights to weigh any integral weight from 1 to 100 kg (when weights can be placed on either pan) is 1 kg, 3 kg, 9 kg, 27 kg and any one integral weight from 60 kg to 81 kg.

Comments: (a) It can be seen that, the set of weights required in this case are successive powers of two i.e., 1 kg (2^0), 2 kg. (2^1), 4 kg (2^2), 8 kg (2^3), 16 kg (2^4) and so on.

To weigh any object weighing in whole number of kg up to, say N kg, only n set of weights are sufficient, where $N < 2^n$ as shown in Table 4.18.

Table 4.18 .

n	N
2 (i.e., 1 and 2 kg)	1 + 2 = 3 kg
3 (i.e., 1, 2 and 4 kg)	1 + 2 + 4 = 7 kg
4 (i.e., 1, 2, 4 and 8 kg)	1 + 2 + 4 + 8 = 15 kg and so on

For N = 100 kg, we can see that n = 7 as $100 < 2^7$. Since the set of weights comprises successive powers of 2, it can be termed a binary set of weights.

Because of the binary set of weights, it is easy to identify the set of weights required for weighing N.

Let N = 43, converting N into binary number,

$1 \times 2^5 + 0 \times 2^4 + 1 \times 2^3 + 0 \times 2^2 + 1 \times 2^1 + 1 \times 2^0 = 101011_2$

So set of weights required include 1, 2, 8 and 32 kg to weigh 43 kg (if only one scale pan is used).

This puzzle has been very popular and appeared in various forms being asked in interviews of various companies, like, "If you lock N rupees bills in n boxes, how can you give any requested amount in whole rupees from 1 to N without opening the box". This type of puzzle can be easily solved using the concept given above.

The problem proposed by Bachet in the 17th century was "the determination of the least number of weights that can be used to weigh any integral number of pounds from 1 to 40 inclusive". He gave the following two solutions:

(i)   Weights of 1, 2, 4, 8, 16 and 32 pounds
(ii) Weights of 1, 3, 9 and 27 pounds.

The first solution pertains to when weights can be placed in one scale pan only, whereas the second solution pertains to when weights can be placed in either of the scale pans.

(b) The set of weights required in this case are successive powers of 3 i.e., $3^0 = 1, 3^1 = 3, 3^2 = 9, 3^3 = 27$ and so on. In this case, weights can be placed in either of the scale pans.

To weigh any object weighing in whole numbers up to N kg, only n set of weights is sufficient, where $N \leq (3^n - 1)/2$ as shown in Table 4.19.

Table 4.19 .

n	N
2 (i.e., 1 and 3 kg)	1 + 3 = 4 kg
3 (i.e., 1, 3 and 9 kg)	1 + 3 + 9 = 13 kg
4 (i.e., 1, 3, 9 and 27 kg)	1 + 3 + 9 + 27 = 40 kg and so on

So, for N = 100 kg, five sets of weights are required as,

$N \leq (3^5 - 1)/2$

$\Rightarrow N \leq 121$

Since the set of weights comprises powers of 3, it can be termed as a ternary set of weights. It will be interesting to find how easily you can find the set of weights by writing the required value of N in the ternary system.

65. Match Stick and Squares

(a) By removing the five match sticks marked m as shown in Fig. 4.34, we obtain the Fig. 4.35 which contains only three squares.

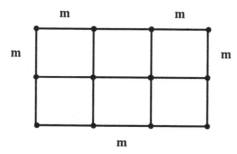

Fig. 4.34 .

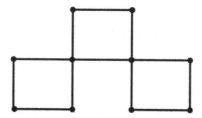

Fig. 4.35 .

(b) By removing the match stick marked X, we obtain the Fig. 4.36 which consists of five squares.

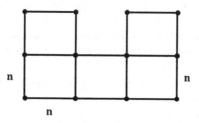

Fig. 4.36 .

By shifting the position of match sticks marked n in Fig. 4.36, we obtain the Fig. 4.37 which consists of exactly four squares.

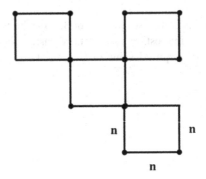

Fig. 4.37 .

66. Insufficient Pounds

There are six different items. Let the cost of these items is a, b, c, d, e and f. Total cost of each of the six combinations of five items is given below:

$$a+b+c+d+e = 33$$
$$a+b+c+d+f = 36$$
$$a+b+c+e+f = 42$$
$$a+b+d+e+f = 43$$
$$a+c+d+e+f = 47$$
$$b+c+d+e+f = 49$$

Adding all the above equations, we get

$$5(a+b+c+d+e+f) = 250$$
$$\Rightarrow a+b+c+d+e+f = 50$$

So, the total cost of all six items is 50 pounds. Since the maximum cost of five items is 49 pounds, which Amit can buy but he is not having sufficient pounds to buy six items as he is short of 50 pounds. Hence Amit was having 49 pounds in his pocket. Since the total cost of six items is known i.e., 50 pounds and the total cost of all combinations of five items is known, so the cost of each item can be easily calculated by subtracting the cost of each combination of the cost of five items from 50 pounds.

The price of each item is $50 - 33 = 17$ pounds, $50 - 36 = 14$ pounds, $50 - 42 = 8$ pounds, $50 - 43 = 7$ pounds, $50 - 47 = 3$ pounds and $50 - 49 = 1$ pound.

67. Mixing Water and Milk

The important point to note is that it is not necessary to know the quantity of liquid in each jar and the quantity transferred. The only key point is that the final volume in both jars is the same as at the beginning. In the end, there is a certain quantity of milk in the water jar. This quantity (volume) of milk must have replaced an equal quantity (volume) of water from the water jar to the milk jar. This equal quantity of water is now available in the milk jar.

Comments: This can be explained in another way by considering grains of wheat and millet instead of water and milk as follows.
Let there are 100 grains of wheat in jar W and 100 grains of millet in jar M. Transfer say 16 grains of wheat from jar W to jar M. Now from jar M, transfer back 16 grains, which is a combination of wheat and millet grains (say 11 grains of millet and 5 grains of wheat) to jar W. Now both jars again contain 100 grains each. Out of these 100 grains,
Number of wheat grains in jar W = 100 − 16 + 5 = 89
Number of millet grains in jar W = 11
Number of wheat grains in jar M = 16 − 5 = 11
Number of millet grains in jar M = 100 − 11 = 89
So, there is as much quantity of millet grains in wheat jar W, as there are wheat grains in millet jar M. This holds good irrespective of initial quantities in each jar and the quantities transferred from one jar to the other subject to the condition that the final quantity in both jars must be the same as at the beginning.

68. The Three Jug—Decanting Liquid

(i) Let the 12, 7 and 5 liters of jugs be denoted by A, B and C respectively as shown in Fig. 4.38.

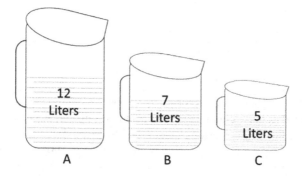

Fig. 4.38 .

Let (a, b, c) denotes the liters of milk in jug A, B and C respectively at any stage. For dividing 12 liters of milk into two equal parts i.e., 6 liters in jug A and 6 liters in jug B, proceed as follows:

There are two ways of pouring milk from a 12-liters jug (i.e., jug A) i.e., either first, fill jug B with 7 liters of milk or fill jug C with 5 liters of milk. These two positions can be denoted as

Case I $(12, 0, 0) \rightarrow (5, 7, 0)$

Case II $(12, 0, 0) \rightarrow (7, 0, 5)$

Where $(12, 0, 0)$ is the initial position with 12 liters of milk in jug A, no milk in jug B and jug C.

Two important aspects to be kept in view are,

(a) To avoid a situation in which jug B and jug C are full because that can be achieved initially itself.

(b) To avoid any position, which has been achieved earlier, as retracing the step will not serve any purpose.

Let's first consider case I, $(12, 0, 0) \rightarrow (5, 7, 0)$.

From $(5, 7, 0)$ two possible positions are $(0, 7, 5)$ or $(5, 2, 5)$ as shown in Fig. 4.39.

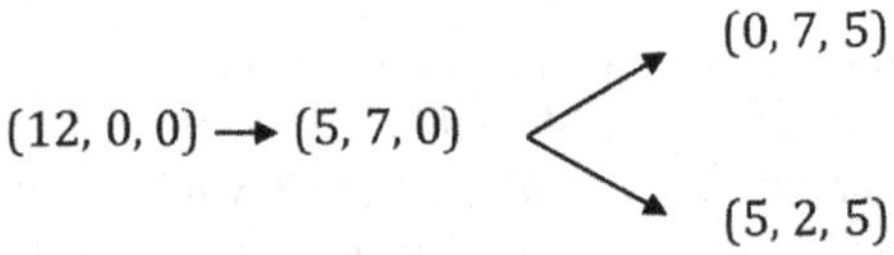

$$(12, 0, 0) \rightarrow (5, 7, 0) < \begin{matrix} (0, 7, 5) \\ (5, 2, 5) \end{matrix}$$

Fig. 4.39 .

As discussed in (a) above, position $(0, 7, 5)$ is to be avoided, so proceed with position $(5, 2, 5)$ only. From $(5, 2, 5)$, four positions are possible as shown in Fig. 4.40.

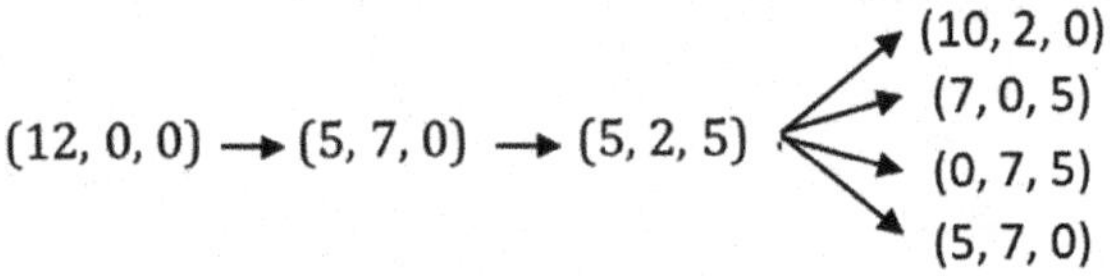

$$(12, 0, 0) \rightarrow (5, 7, 0) \rightarrow (5, 2, 5) < \begin{matrix} (10, 2, 0) \\ (7, 0, 5) \\ (0, 7, 5) \\ (5, 7, 0) \end{matrix}$$

Fig. 4.40 .

Out of four possible positions for $(5, 2, 5)$, positions $(7, 0, 5)$ and $(0, 7, 5)$ are to be discarded as these could have been achieved initially. Position $(5, 7, 0)$ is to be discarded as earlier positions are not to be retraced.

From $(10, 2, 0)$, again three positions are possible as shown in Fig. 4.41.

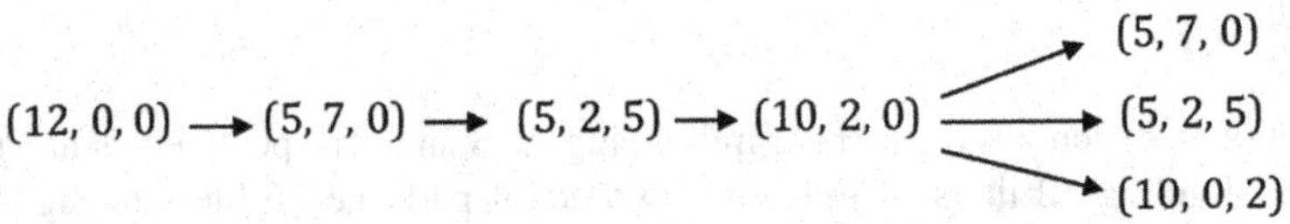

$$(12, 0, 0) \rightarrow (5, 7, 0) \rightarrow (5, 2, 5) \rightarrow (10, 2, 0) < \begin{matrix} (5, 7, 0) \\ (5, 2, 5) \\ (10, 0, 2) \end{matrix}$$

Fig. 4.41 .

Out of three possible positions from (10, 2, 0), positions (5, 7, 0) and (5, 2, 5) are to be discarded as earlier positions are not to be retraced. By proceeding in this way, the following solution can be obtained in 11 decantings.

(12, 0, 0) → (5, 7, 0) → (5, 2, 5) → (10, 2, 0) → (10, 0, 2) → (3, 7, 2) → (3, 4, 5) → (8, 4, 0) → (8, 0, 4) → (1, 7, 4) → (1, 6, 5) → (6, 6, 0)

Similarly, we can start with case II i.e., (7, 0, 5) and proceeding as above, we find that minimum 12 decantings are required in this case as shown below:

(12, 0, 0) → (7, 0, 5) → (7, 5, 0) → (2, 5, 5) → (2, 7, 5) → (9, 0, 3) → (9, 3, 0) → (4, 3, 5) → (4, 7, 1) → (11, 0, 1) → (11, 1, 0) → (6, 1, 5) → (6, 6, 0)

So, the least number of decantings required is eleven, to divide the milk into two equal parts by first starting decanting in jug B as per case I.

(ii) This part of the puzzle can also be solved similarly as described above. However, let's solve it in a different way using graph paper. Consider a graph paper of parallelogram shape, with sides as 7 units and 5 units, divided into triangles as shown in Fig. 4.42.

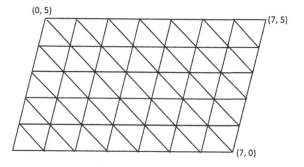

Fig. 4.42 .

The coordinates of any point can easily be calculated as shown in Fig. 4.43. The x-coordinate is along the horizontal axis, whereas the y-coordinate is along the inclined side of the parallelogram.

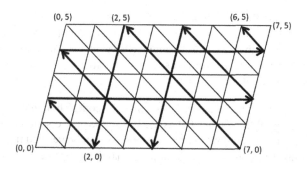

Fig. 4.43 .

As in the previous part of the puzzle, we can have two options i.e., either first fill 7 liters jug or 5 liters jug. Let's consider that the 7 liters jug is filled first. So, the initial position (a, b) is (7, 0), where a is the content of 7 liters jug and b is the content of 5 liters jug. Start with the bottom right corner whose coordinates are (7, 0) as shown in Fig. 4.43.

From (7, 0), proceed in the direction of the line of the triangle till it meets the edge of the graph paper i.e. (2, 5).

From this position (2, 5), again proceed in the direction of the line of the triangle without retracing the path, till it meets the edge i.e., at (2, 0). Continue in this way till we meet the coordinate (6, 5).

It can be seen that, starting from (7, 0) and traversing the coordinates (2, 5), (2, 0), (0, 2), (7, 2), (4, 5), (4, 0), (0, 4) and (7, 4), we reach to final destination i.e. (6, 5).

So, 6 liters of milk can be supplied in ten decantings as given by the coordinates shown in Fig. 4.43 i.e.

$(0, 0) \rightarrow (7, 0) \rightarrow (2, 5) \rightarrow (2, 0) \rightarrow (0, 2) \rightarrow (7, 2) \rightarrow (4, 5) \rightarrow (4, 0) \rightarrow (0, 4) \rightarrow (7, 4) \rightarrow (6, 5)$

If we start by filling the 5 liters jug first and proceeding in the same way as given above, 12 decantings are required as under:

$(0, 0) \rightarrow (0, 5) \rightarrow (5, 0) \rightarrow (5, 5) \rightarrow (7, 3) \rightarrow (0, 3) \rightarrow (3, 0) \rightarrow (3, 5) \rightarrow (7, 1) \rightarrow (0, 1) \rightarrow (1, 0) \rightarrow (1, 5) \rightarrow (6, 0)$

69. Crack the 3-Digit Lock Code

(i)   Clues A and B indicate that 5 is at the same place, so 5 cannot be a correct number, as it cannot be at the right and wrong place at the same time.

(ii) From clue D, only two numbers are correct, so the correct numbers are 3 and 0, as 5 is not a correct number, as already deduced from clues A and B above.

(iii) From clue D, two numbers 3 and 0 are correct as deduced above but both these numbers are incorrectly placed. Since 3 is correctly placed at the last position (from clue A), obviously correct position of '0' is in the first place.

(iv) From clue B, one number is correct but wrongly placed. This number can either be 1 or 8. Since 1 is placed in the middle position, it cannot be a correct number, as it is wrongly placed in the middle position, so the correct number is 8 and its correct position is in middle. Hence the correct code is $\boxed{0 \mid 8 \mid 3}$

Comments: There are 5 clues given to crack the code. Though it is easy to crack the code using all the 5 clues but as seen from the above solution, only 3 clues i.e. A, B and D are required to crack the code. It can be verified that the solution also satisfies the conditions given in the remaining clues i.e., C and E also.

70. Gold and Silver Bricks

There are three gold bricks and two silver bricks, out of which, any three bricks are contained in three lockers E, M and Y.

(i) When the eldest son, after seeing the content of locker M and Y says, he does not know the content of his locker E, this conveys that locker M and Y both cannot contain silver bricks because in that case, the content of locker E will certainly be gold brick, which could have been easily deduced by the eldest son. So, both lockers M and Y, either contain gold bricks or one contains gold brick and the other silver brick.

(ii) Middle son knows the content of locker Y and after hearing the statement of elder son says, he does not know the content of his locker M. This conveys that locker Y cannot contain silver brick. Because if Y contains silver brick, then as deduced from (i) above, Y and M both cannot contain silver brick, so M certainly contains gold brick, which he can easily deduce. Hence Y cannot contain silver brick. This fact is now known to the youngest son, who says with certainty that he knows the content of his locker Y i.e., gold brick.

71. Three Men and Their Wives

This is an excellent puzzle requiring the use of logic and arithmetic.

Since each person buys as many apples as he or she pays for each apple in rupees, so the amount spent by each person is a perfect square.

Each man spends 45 rupees more than his wife, so the difference in expenditure between each couple is 45. Number 45 can be expressed as a product of two numbers in three different ways as follows:

$45 = 1 \times 45 = 3 \times 15 = 5 \times 9$

So, 45 can be expressed as the difference of two squares in three different ways, $23^2 - 22^2 = 45, 9^2 - 6^2 = 45$ and $7^2 - 2^2 = 45$.

It can be implied that 23^2, 9^2 and 7^2 represent expenditure by men and 22^2, 6^2 and 2^2 represent expenditure by women.

Since A buys 17 more apples than Y, so A spends 23^2 and Y spends 6^2, as $23 - 6 = 17$. Similarly, B spends 9^2 and Z spends 2^2, as $9 - 2 = 7$.

So, X is the wife of A, Y is the wife of B and Z is the wife of C.

Comments: A number can be expressed as the difference of two squares in as many ways as it can be expressed as the product of two odd or two even numbers. The only exceptions are 1, 4 and twice any odd number.

72. Hundred Balloons in Hundred Rupees

Let number of green balloons, yellow balloons and red balloons are x, y and z respectively, so

$$x + y + z = 100 \tag{i}$$

$$5x + y + z/20 = 100 \tag{ii}$$

Subtracting Eqs. (i) from (ii), we get $4x = \frac{19z}{20}$,

$\Rightarrow z = \frac{80X}{19}$

For z to be an integer, x must be a multiple of 19.

If x = 19, z = 80 and y = 1.

If x = 38 or higher multiple of 19, z > 100 which violates condition of Eq. (i).

Hence Rishik buys 19 green, 1 yellow and 80 red balloons for 100 rupees.

73. **Counterfeit Coin**

Divide the 12 coins into three groups of four coins each. Let the coins in these three groups be denoted by ABCD, EFGH and IJKL.

Balance any two groups say ABCD and EFGH.

(i) If ABCD = EFGH, it implied that the defective coin is in third group IJKL.

In the second balancing, place three coins of the defective group, say IJK on one side of the scale and any three good coins (from two groups of good coins i.e., ABCD and EFGH) say ABC.

a) If IJK = ABC, then the remaining fourth coin of the defective group i.e., coin L is defective.

In the third balancing, balance coin L with any good coin to know whether defective coin L is lighter or heavier.

b) If IJK > ABC, then the defective coin is I, J or K and it is heavier than a good coin.

In third balancing, balance any two of the I, J or K coin.
Say I and J. If I > J, then I is defective and heavier.
If I < J, then J is defective and heavier.
If I = J, then K is defective and heavier.

c) If IJK < ABC, then the defective coin is I, J or K and it is lighter than a good coin.

In third balancing, balance any two of I, J or K say I and J.
If I < J, then I is defective and lighter.
If I > J, then J is defective and lighter.
If I = J, then K is defective and lighter.

(ii) If ABCD > EFGH, it implied that all four coins in the third group i.e. I, J, K and L are good coins. Also, if the defective coin is in group ABCD, it is heavier than a good coin. If the defective coin is in group EFGH, it is lighter than a good coin.

In the second balancing, take three coins A, B and E on one side of the scale and another three coins C, D and F on the other side of the scale.

a) If ABE = CDF, then either G or H is defective and is lighter than a good coin.

In third balancing, balance G and H,
If G > H, then coin H is defective and lighter.
If G < H, then coin G is defective and lighter.

b) If ABE > CDF, then coins C, D and E are good coins because if C and D were heavier or E is lighter then, CDF > ABE. So, either coin A or B is a heavier defective coin or coin F is lighter and defective.

In third balancing, balance A and B.
If A = B, then F is a defective coin and lighter.
If A > B, then A is a defective coin and heavier.
If A < B, then B is a defective coin and heavier.

c) If ABE < CDF, as explained above, in this case, A, B and F are good coins.

In third balancing, balance C and D,
If C = D, then E is a defective coin and lighter.
If C > D, then C is a defective coin and heavier.
If C < D, then D is a defective coin and heavier.

(iii) If ABCD < EFGH

Based on similar logic as given in (ii), we can identify the defective coin and its defect whether lighter or heavier in three balancings.

Comments: For N coins, where defect whether lighter or heavier is also to be identified, then minimum balancings n required are given by

$$N = (3^n - 3)/2$$

For the given puzzle N = 12, so, n = 3 balancings are required.
This is a classic puzzle and has appeared in literature in many places for a long time.
There can be many other solutions to this puzzle in three balancings. One of the other solutions is briefly described below:
Divide the 12 coins into three groups of four coins each. In first balancing, balance any two groups as done in the earlier solution. If the scale balances, the defective coin lies among the third group and can be identified in a total of three balancings as done earlier.
Suppose the two groups of four coins do not balance in the first balancing i.e., one group is heavier than the other, then the four coins in the third group are good coins. Let's denote the coins in the heavier group as H1, H2, H3 and H4, coins in the lighter group as L1, L2, L3 and L4, and good coins as G1, G2, G3 and G4.

In first balancing, let H1H2H3H4 > L1L2L3L4, this indicates that G1, G2, G3 and G4 are good coins. If the defective coin is from the H1H2H3H4 group, then it is heavier else lighter.

In the second balancing, select groups of four coins as H1H2H3L1, H4G1G2G3 and L2L3L4G4.

(i) If H1H2H3L1 = H4G1G2G3, then since G4 is a good coin, so L2 or L3 or L4 is defective and lighter. In the third balancing, balance L2 and L3.

If L2 > L3, then L3 is defective and lighter.
If L2 < L3, then L2 is defective and lighter.
If L2 = L3, then L4 is defective and lighter.

(ii) If H1H2H3L1 > H4G1G2G3, then H1 or H2 or H3 is defective and heavier.

In the third balancing, balance H1 and H2.
If H1 > H2, then H1 is defective and heavier.
If H1 < H2, then H2 is defective and heavier.
If H1 = H2, then H3 is defective and heavier.

(iii) If H1H2H3L1 < H4G1G2G3, then either L1 is defective and lighter or H4 is defective and heavier.

In the third balancing, balance either L1 or H4 against a good coin say G1.
If L1 = G1, then H4 is defective and heavier.
If L1 < G1, then L1 is defective and lighter.
This way we can identify the defective coin along with its defect whether lighter or heavier in three balancings.

74. Effect of Paper Folding

The thickness of paper = 0.1 mm
After each fold, the thickness of the paper becomes two times, so after n folds, thickness will be = 0.1×2^n mm. After 10 folds, n = 10, thickness = 0.1×2^{10} = 102.4 mm.
One side of the paper becomes half after each fold, so both sides of the paper will become half after two folds. After 10 folds, each side of paper = $1000/2^5$ = 31.25 mm.
So, the thickness of paper becomes more than three times the width and length of paper, which makes it so thick that it may not be possible to fold it further.
After 20 folds, thickness = 0.1×2^{20} mm = 104.8576 m
After 30 folds, thickness = 0.1×2^{30} mm = 107.37 km
Can you believe that a paper with 0.1 mm thickness after 30 folds becomes more than 107 km thick?
With 50 folds, thickness = 0.1×2^{50} mm $\simeq$ 112589990 km
This is about three fourth of the distance from the earth to the sun, which cannot be thought off.
On similar lines, the size of paper after 20, 30 and 50 folds can be calculated.

Comments: Exponential growth makes results highly unexpected in such types of puzzles.

You may be aware of an old legend about the reward for inventing chess. According to legend, chess was invented by Sissa Ben Dahir. King Shirham of India was so delighted that he offered him any reward he requested.

Sissa said, "Give me just one grain of wheat on the first square of the chess board, two grains of wheat on the second square, four grains on the third square, eight grains on the fourth square and continue doubling the number of grains on each successive square until every square on the chessboard is covered". King laughed at Sissa for asking for such a small reward. King was surprised when he came to know that Sissa asked for more wheat than he has in his entire kingdom and even more than the production of wheat in the entire world for decades.

The total number of grains of wheat

$$= 1 + 2^1 + 2^2 + 2^3 + \cdots + 2^{63}$$
$$= 2^{64} - 1$$

Assuming about 16 grains of wheat in one gram, one kg will contain 16000 grains.

$$\text{Weight of } 2^{64} - 1 \text{ grains} = \frac{2^{64} - 1}{16000} \text{ kg}$$
$$= 10^{15} \text{ kg} = 10^{12} \text{ tonnes}$$

75. Seven Security Posts

For the circular perimeter of 13 km, four security posts can be placed with distances between adjoining posts as 1, 2, 6 and 4 km on the circular path such that all integral distances from 1 to 12 km can be covered between two security posts as given in Table 4.20.

Table 4.20 .

Distance (km)	Clockwise measurement between two security posts
1	A to B
2	B to C
3	A to C
4	D to A
5	D to B
6	C to D
7	D to C
8	B to D
9	A to D
10	C to A
11	C to B
12	B to A

Another solution for four security posts is given by inter distances as 1, 3, 2 and 7 km. Other solutions can be obtained by translation.

For a 39 km perimeter, seven security posts can be placed at 0, 1, 2, 4, 13, 18 and 33 km respectively. The inter distance between two adjacent security posts will be 1, 1, 2, 9, 5, 15 and 6. With these positions of security posts, all integral distances from 1 km to 38 km can be covered between any two security posts. Other solutions can also be obtained.

If the number of security posts is n, then the number of possible trips between different security posts = n (n − 1).

In addition, one trip consisting of a complete round can be made from any security post, so for n security posts, the maximum number of trips.

$$= n(n − 1) + 1$$
$$= n^2 − n + 1$$

For n = 7, maximum number of trips = $7 \times 7 − 7 + 1 = 43$. Since the perimeter is only 39 km, this means four lengths will be repeated.

Comments: For various ranges of the length of circular perimeter up to 39 km, the minimum number of security posts required to cover all integral distances from one security post to the other has been computed. For example, five security posts are sufficient for trips of all integral lengths from 1 to 21 km.

For n = 5, maximum number of possible trips = $n^2 − n + 1 = 21$

So, for n = 5, all possible maximum trips i.e., 21 are possible. This is an optimum solution. The posts can be located at 0, 1, 4, 14 and 16 km, with inter distances between adjacent posts as 1, 3, 10, 2 and 5 km. Other solutions for n = 5 can be obtained by translation.

For n = 6, maximum number of trips = $6 \times 6 − 6 + 1 = 31$

This also gives optimum solution as posts can be located at 0, 1, 3, 8, 12 and 18 km, with inter distances between adjacent posts as 1, 2, 5, 4, 6 and 13 km. Many other solutions are possible for n = 6, some of which are:

(i)   0, 1, 3, 10, 14 and 26 km, with inter distances between adjacent posts as 1, 2, 7, 4, 12 and 5 km.
(ii) 0, 1, 4, 6, 13 and 21 km, with inter distances between adjacent posts as 1, 3, 2, 7, 8 and 10 km.
(iii) 0, 1, 4, 10, 12 and 17 km, with inter distances between adjacent posts as 1, 3, 6, 2, 5 and 14 km.

It can be seen that for n ≤ 6, an optimum solution can be obtained, so there is no repetition of trip lengths. However, for n > 6, there may be a repetition of trip length as already seen for n = 7. The details are shown in Table 4.21.

Table 4.21 .

Number of security post n	Range of perimeter (km)	$n^2 - n + 1$
1	0–1	1
2	2–3	3
3	4–7	7
4	8–13	13
5	14–21	21
6	22–31	31
7	32–39	43

76. Rotating Balls

Let total number of balls = x and number of black balls = y

So, $\frac{75x}{100} < y < \frac{80x}{100}$

$\Rightarrow \frac{3x}{4} < y < \frac{4x}{5}$

$\Rightarrow y > \frac{3x}{4} \Rightarrow \frac{x}{y} < \frac{4}{3}$ and $y < \frac{4x}{5} \Rightarrow \frac{x}{y} > \frac{5}{4}$

For positive fractions $\frac{a}{b}$ in which $\frac{a}{b} < \frac{c}{d}$

$\frac{a}{b} < \frac{a+c}{b+d} < \frac{c}{d}$

Since $\frac{5}{4} < \frac{x}{y} < \frac{4}{3} \Rightarrow \frac{5}{4} < \frac{5+4}{4+3} < \frac{4}{3}$

So, $x = 9$ and $y = 7$

So total number of balls = 9 and number of black balls = 7.

77. Holidays During Covid-19 Crisis

Let the number of holidays = x days and number of office working days = y days.

Each day, he goes for 6 km walk during x days of holidays.

Let each day he goes for N km walk during y days of office working.

Average period of walk during the whole lockdown period is

$(6x + N y)/(x + y) = 3N$

$\Rightarrow 6x + Ny = 3N(x + y)$

$\Rightarrow x(6 - 3N) = 2Ny$

$\Rightarrow 3x(2 - N) = 2Ny$

$\Rightarrow N = 1$ and $\frac{x}{y} = \frac{2}{3}$

So, x + y must be a multiple of 5.

Now x + y = 8 to 9 weeks (Given)

$$= 56 \text{ to } 63 \text{ days.}$$

Hence $x + y = 60$ (As 60 is the only multiple of 5 in the range 56–63 days)
$\Rightarrow x = 24$ and $y = 36$
So, he enjoyed 24 days of holiday during the lockdown period.

78. **Making the Chain from Cut Pieces**

(i) Before attempting a solution to the given puzzle, let's consider a simple case
of five pieces of chain, each consisting of four links as shown in Fig. 4.44.

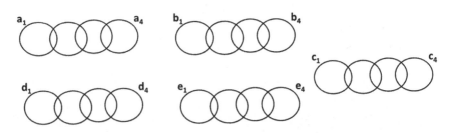

Fig. 4.44 .

The simplest way to connect these five pieces into an endless i.e., circular chain
is by cutting one end link of each piece (i.e., five cuts at a_4, b_4, c_4, d_4 and e_4 in
five links) and joining. This involves five cuts and five solderings, resulting in a
total cost of $5 \times 5 = 25$ dollars. Instead of it, if cuts are made in all four links
of one piece (i.e. e_1, e_2, e_3 and e_4) and joining the remaining four pieces by
these four-cut links (i.e., a_4–e_1–b_1, b_4–e_2–c_1, c_4–e_3–d_1 and d_4–e_4–a_1), an endless chain of 20 links can be made. This involves only four cuts and four
soldering's, thus reducing the total cost to $4 \times 5 = 20$ dollars and is the
minimum possible cost. Hence cutting end links of every piece and then joining
may not be an optimum solution.
The main objective is to minimize the number of cuts to obtain the minimum
cost.
On the same analogy as discussed earlier, if we have a piece of nine links, these
could have been cut into nine single links to join nine ends of the remaining
nine pieces into an endless (circular) chain. But nine link pieces is not available.
Similarly, if we have two pieces containing a total of eight links, these can be
cut into eight single links to join eight ends of the remaining eight pieces into
an endless chain. This is possible by cutting one five-link piece and one
three-link piece. The total cost comes to $8 \times 5 = 40$ dollars.
However, we can find that three pieces containing a total of seven links are
available (i.e., 3 link piece + 2 link piece + 2 link piece), which can be used to
cut into seven single links and join the remaining seven pieces into an endless
chain. This reduces the total cost to $7 \times 5 = 35$ dollars only.

We don't have four pieces of a total of six links, so the cost as obtained above i.e., 35 dollars is the minimum.

(ii) Before attempting the solution to the given puzzle, let's consider a simple case of five pieces of chain, each consisting of 3 links as shown in Fig. 4.45.

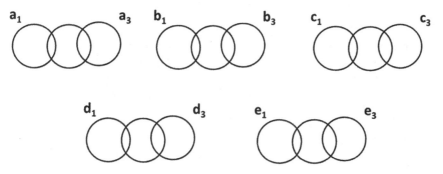

Fig. 4.45 .

As explained in part i) of the puzzle, instead of cutting the ends of four pieces i.e., a_3, b_3, c_3 and d_3, cut all three links of one piece and join at the ends of three pieces i.e., at a_3, b_3 and c_3 to get a long open-ended chain of 15 links for $3 \times 5 = 15$ dollars only.

On the same analogy as discussed in part i) of the puzzle, instead of cutting end links of nine pieces and joining them to get an open-ended chain of 40 links, it is possible to cut three pieces totalling six links (i.e., 2 link piece + 2 link piece + 2 link piece) into six single links and joining the remaining pieces as discussed earlier, to get an open-ended long chain of 40 links. The cost, in this case, is reduced to $6 \times 5 = 30$ dollars only. Since four pieces containing five links in total are not available, the cost of 30 dollars obtained above is the minimum cost.

Comments: If we have n pieces of chain, each having $n - 1$ links, the minimum cost is obtained by cutting all links i.e. $(n - 1)$ links of one piece and joining the remaining $(n - 1)$ pieces into an endless chain of $n \times (n - 1)$ links. Similarly, if we have n pieces of chain, each having $n - 2$ links, the minimum cost is obtained by cutting all links i.e., $n - 2$ links of one piece and joining the remaining pieces into an open-ended chain of $n \times (n - 2)$ links.

79. Four Operations on Two Numbers

The sum of four operations = 243

$$\Rightarrow (x+y)+(x-y)+xy+x/y = 243$$
$$\Rightarrow 2x+xy+x/y = 243$$
$$\Rightarrow x(2y+y^2+1) = 243y$$
$$\Rightarrow x = \frac{243y}{(y+1)^2}$$

Since y cannot have any common factor with y + 1 (because two consecutive numbers are always co-prime), $(y+1)^2$ must be a divisor of 243 to ensure x as a whole number.

Since $243 = 3^5$, so $(y+1)^2$ can be either equal to 3^2 or 3^4.

If $(y+1)^2 = 3^2 \Rightarrow y = 2$

If $(y+1)^2 = 3^4 \Rightarrow y = 8$

If $y = 2$, $x = \frac{243 \times 2}{3^2} = 54$

If $y = 8$, $x = \frac{243 \times 8}{9^2} = 24$

So, the two solutions of required numbers are

(i) x = 24 and y = 8

(ii) x = 54 and y = 2

80. Who'll Become a Decamillionaire (Kaun Banega Crorepati)?

To solve such puzzles, one should start from the end and go backward.

To win, one must get the resultant product equal to or more than 10^7 i.e., one decamillion (one crore).

Suppose one gets the value of the product as x from his opponent, then to win, the value of $x = x_w$ can be computed as under:

$$x_w \times 2 \geq 10^7, \text{ so } x_w \geq 5000000$$
$$x_w \times 10 \geq 10^7, \text{ so } x_w \geq 1000000$$

So, the value of x_w can be 1000000 or higher up to $10^7 - 1$, which on multiplication by a chosen multiplier from 2 to 10, gives the desired resultant product i.e., 10^7 or more. By proceeding backward in a similar way, a further range of values $x = x_L$ can be obtained, which on multiplication by the chosen multiplier from 2 to 10 results in losing position of the opponent.

Since minimum $x_w = 1000000$, so

$$\text{Minimum } x_L \times 2 \geq \text{Minimum } x_w$$
$$\Rightarrow x_L \geq \frac{1000000}{2} = 500000$$

and maximum value of $x_L = \text{Minimum } x_w - 1 = 999999$

Similar process can be repeated to obtain minimum values of x_w and x_L at every stage till we reach last but one stage as given below:

	Range
x_w	1000000–9999999
x_L	500000–999999
x_w	50000–499999
x_L	25000–49999
x_w	2500–24999
x_L	1250–2499
x_w	125–1249
x_L	63–124

For x_w (min.) = 125, $x_L \geq \frac{x_w}{2}$ (i.e. 62.5), so minimum x_L = 63 being an integer value.

The next minimum value of x_w can be calculated as

$10x_w \geq 63 \Rightarrow x_w \geq \frac{63}{10}$ (i.e., 6.3), so minimum x_w = 7 being an integer value and maximum x_w = 63 − 1 = 62

For the last stage, we compute the last value of x_L (which is the first value of x_L).

$2x_L \geq 7 \Rightarrow x_L \geq \frac{7}{2} = 3.5$, so minimum x_L = 4 and maximum x_L = 7 − 1 = 6

Hence the opponent person B must be given a product value from 4 to 6 by person A, to ensure a win by person A. That is possible if person A multiplies his initial number 1 given to him by 4, 5 or 6 only.

Other subsequent values of x_L and x_w in continuation to previous data are given below:

	Range
x_L	63–124
x_w	7–62
x_L	4–6
x_w	1

It may be noted that to find the minimum value of x_L (backward), the minimum value of x_w is divided by the smallest multiplier i.e., 2 (rounded off to the next integer) and to find the minimum value of x_w (backward), the minimum value of x_L is divided by the largest multiplier i.e., 10 (rounded off to the next integer) except in the last step, where the value of x_w = 1 is already given to start the game. It can be seen that the person who wins the toss is given the number 1 i.e., person A and he finally ensures his win by forcing products x_L (losing products) to opponent person B such that whatever multiplier is used by opponent person B, he cannot win except when person A himself makes a mistake at any of the steps.

81. India's Independence Day

There are seven days in a week, so any month having 31 days like August, will have three extra weekdays and these will fall five times in that month. The other four weekdays will fall four times in that month. Obviously, the 1st, 2nd and 3rd of the month will have weekdays which occur five times in the month.

Since Monday and Thursday fall four times in August (given), Tuesday and Wednesday will also fall four times in August, as these days are sandwiched between Monday and Thursday. So, the 4th of the month must fall on Monday. Hence 15th August i.e., India's Independence Day will fall on Friday in that year.

82. Palindromic Mileage

$$\text{Average speed} = \frac{\text{Total Distance Travelled}}{\text{Total Time}}$$

$$\text{Total time given} = 11\!:\!58\ \text{AM} - 10\!:\!18\ \text{AM}$$
$$= 100\ \text{minutes}$$

Total distance travelled in 100 minutes

= Difference in two milometer readings, which are palindromic

number 78987 and the next palindromic number.

Next palindromic number is obviously 79097, so total distance travelled = 79097 − 78987 = 110 km.
So average speed = $\frac{110}{100}$ = 1.1 km per minute = 66 km per hour.

83. Time Measurement with Hourglass

An hourglass is a device comprised of two glass bulbs connected vertically by a narrow tube that allows a regulated flow of sand from the upper bulb to the lower bulb as shown in Fig. 4.46.

Fig. 4.46 .

A 4-minute hourglass means, the complete sand from the upper bulb will fall into the lower bulb in 4 minutes.

There can be many solutions to measure nine minutes but the fastest way in which nine minutes can be measured is given below:

(i) Start both hourglasses at zero minutes.

(ii) After 4 minutes, flip the 4-minute hourglass. At this point, 3 minutes of sand is left in 7-minute hourglass.

(iii) After 7 minutes from start, flip the 7-minute hourglass. At this point, 1 minute of sand is left in 4-minute hourglass.

(iv) After 8 minutes from start, 4-minute hourglass is emptied. At this point 6 minutes of sand is left in 7-minute hourglass i.e., 1 minute of sand is fallen in the lower bulb. So, we can flip 7-minute hourglass at this point to measure 1 minute more. Hence at 9 minutes from start, the 7-minute hourglass is also emptied (upper bulb after the flip). At this point, the time will be 9 minutes from start.

The details explained above are summarized in Table 4.22.

Table 4.22 .

Timing →	4 minutes	7 minutes	8 minutes	9 minutes
4-minute hourglass	Flip (empty)	1 min. sand is left	Empty	
7-minute hourglass	3 min. sand is left	Flip (empty)	1 min. sand is left flip	Empty
	←	9 minutes	→	

Comments: Another variation of this puzzle is to find the quickest way to measure 15 minutes from 7-minute hourglass and 11-minute hourglass. The solution is briefly given below:

(i) Start both hourglasses at zero minutes.

(ii) After 7 minutes, flip the empty 7-minute hourglass. At this point, 4 minutes of sand is left in 11-minute hourglass.

(iii) After 11 minutes from start, 3 minutes of sand is left in the upper bulb and 4 minutes of sand is fallen in the lower bulb of 7-minute hourglass. Since we need only 4 minutes, so flip the 7-minute hourglass, which will get emptied in 4 minutes after completion of a total of 15 minutes.

84. Smallest Product of Two Numbers

For the smallest possible product, the first digit of both numbers should be the smallest. Since zero cannot be the first digit, so the first digit of one number will be 1 and the other number will be 2.

Let the numbers are ABCDE and FGHIJ,

So let $A = 1$ and $F = 2$

The choice for the second digit in both numbers is 0 and 3.

(i) If B = 0, then G = 3,
    AB = 10, FG = 23 and FG − AB = 13

(ii) If B = 3, then G = 0,
    AB = 13, FG = 20 and FG − AB = 7

For smallest possible product, the difference of two numbers must be largest for the same sum. Comparing (i) and (ii) for FG − AB, it can be seen that 13 > 7, so B = 0 and G = 3. Proceeding in similar way, the remaining digits can be placed alternatively as follows:

AB = 10, FG = 23
ABC = 104, FGH = 235 [as (235 − 104) > (234 − 105)]
ABCD = 1046, FGHI = 2357 [as (2357 − 1046) > (2356 − 1047)]
ABCDE = 10468, FGHIJ = 23579 [as (23579 − 10468) > (23578 − 10469)]
So, 10468 × 23579 = 246824972 is the smallest possible product.

Comments: A shortcut to the above solution is obtained by placing the digits in ascending order, keeping in view the position of zero which can be placed, after the first digit of either of the two numbers.

$$1, 2, 0, 3, 4, 5, 6, 7, 8, 9$$

Form two numbers taking alternate digits, so the first number is 10468 and the second number is 23579 and the smallest possible product is

$$10468 \times 23579 = 246824972$$

Instead of 10 digits, if two numbers are to be formed from all nine digits i.e., 1 to 9, to obtain the smallest possible product. Proceed in the same way and place all digits 1 to 9 in ascending order as 1, 2, 3, 4, 5, 6, 7, 8 and 9.

The first four digits of each number are formed by taking alternate digits i.e., 1357 and 2468.

Now the last digit 9 shall be placed as the last digit of the number whose leading digit is larger, as this will give the smallest product.

Also 24689 − 1357 > 13579 − 2468

So, the two numbers will be 1357 and 24689. The smallest possible product, in this case, will be 1357 × 24689 = 33502973.

85. Squares and Rectangles on a Chessboard

(i) The smallest square is obviously of size 1 × 1 and the largest size is 8 × 8. Squares of other sizes between 1 × 1 and 8 × 8 also exist.

It can be seen that there are 8 × 8 = 64 squares of size 1 × 1. Similarly, number of squares of size 2 × 2 is 49 i.e., 7^2. Number of squares of size 3 × 3 is 36 i.e., 6^2, number of squares of size 4 × 4 is 25 i.e., 5^2, number of squares of size 5 × 5 is 16 i.e., 4^2, number of squares of size 6 × 6 is 9 i.e., 3^2, number of squares of size 7×7 is 4 i.e., 2^2 and number of squares of size 8×8 is 1 i.e., 1^2.

So total number of squares $= 1^2 + 2^2 + 3^2 + 4^2 + \cdots + 8^2$

$$= \frac{8 \times (8+1)(2 \times 8 + 1)}{6} = 204$$

This can be generalized to $n \times n$ board in which the number of squares is obtained by summing the series of first n squares i.e.

$$= 1^2 + 2^2 + 3^2 + \cdots + n^2$$

$$= \frac{n(n+1)(2n+1)}{6}$$

(ii) On the 8×8 chessboards, there are nine vertical lines and nine horizontal lines. A rectangle including a square can be formed by selecting any two vertical lines from nine vertical lines and any two horizontal lines from nine horizontal lines.

Any two lines can be selected out of 9 lines in $^9C_2$ ways.
So total number of rectangles including squares can be formed in

$$^9C_2 \times {}^9C_2 = 1296$$

Number of squares formed = 204 (as calculated in part (i) of the puzzle).
So, number of rectangles excluding squares which can be formed = 1296 − 204 = 1092.
If the size of board is $n \times n$, then
Total number of rectangles including squares formed $= {}^{(n+1)}C_2 \times {}^{(n+1)}C_2$

$$= \frac{n(n+1)}{2} \times \frac{n(n+1)}{2} = \left[\frac{n(n+1)}{2}\right]^2 = 1^3 + 2^3 + 3^3 + \cdots + n^3$$

Interestingly this is also equivalent to the square of the nth triangular number.
So, the number of rectangles excluding squares.

$$= \left[\frac{n(n+1)}{2}\right]^2 - \frac{n(n+1)(2n+1)}{6}$$

Comments: If the size of the board is $m \times n$, then the number of squares and rectangles formed can be calculated similarly.
A rectangle can be formed by selecting two lines from $(m + 1)$ lines and two lines from $(n + 1)$ lines. So total rectangles formed $= {}^{(m+1)}C_2 \times {}^{(n+1)}C_2$

$$= \frac{m(m+1)}{2} \times \frac{n(n+1)}{2}$$

Total number of squares, which can be formed from m × n board, can be obtained as follows:

For 1 × 1 square, number of squares = m × n

For 2 × 2 squares, number of squares = (m − 1) × (n − 1)

Let m > n, for n × n square (largest possible),

Number of squares = [m − (n − 1)] × [n − (n − 1)] = m − n + 1

On summing all squares i.e., m × n + (m − 1) (n − 1) + ..., we can obtain the number of squares in the m × n board.

86. New Year Party and the Poisoned Drink

Since limited time is available, we cannot wait for the first rat to die and then try the second rat and so on. The required number of rats is to be fed in parallel and see which rats are died to identify the poisoned bottle. Also, note that one rat can be fed from as many bottles as required.

To solve such puzzles, it is always better to start with a small sample and look for a pattern.

So, let's start with two bottles i.e., B1 and B2. One rat is sufficient to identify the poisoned bottle as the only rat will be fed from one bottle say B1 and wait up to 12 hours, if it dies then B1 is the poisoned bottle else B2 is the poisoned bottle.

Now let's consider four bottles i.e., B1, B2, B3 and B4. Try 2 rats to identify the poisoned bottle.

The first rat is fed from bottles B1 and B2. The second rat is fed from bottles B2 and B3.

If only the first rat dies, B1 is a poisoned bottle.

If only the second rat dies, B3 is a poisoned bottle.

If both rats die, B2 is a poisoned bottle.

If no rat dies, B4 is a poisoned bottle.

Similarly, we can consider 8 bottles and will find that 3 rats are sufficient to identify the poisoned bottle. The key point is how to identify bottles in a unique way for a given set of rats. We have seen that for $2^1 = 2$, $2^2 = 4$ and $2^3 = 8$ bottles, one, two and three rats respectively are sufficient to identify the poisoned bottle. So, a pattern is seen.

In such types of puzzles, the use of the binary system of numbers is very useful. Consider the number of rats as the number of bits. So, 3 bits for 3 rats are sufficient to uniquely identify $2^3 = 8$ bottles. Table 4.23 shows 8 bottle numbers in decimal as well as a binary system along with the corresponding set of rats.

Table 4.23 .

Bottle numbers		Corresponding set of rats	Remarks
Decimal system	Binary system		
1	001	{1}	Rat 1 is fed from bottle 1
2	010	{2}	Rat 2 is fed from bottle 2
3	011	{1, 2}	Rat 1 & 2 fed from bottle 3
4	100	{3}	Rat 3 is fed from bottle 4
5	101	{1, 3}	Rat 1 & 3 fed from bottle 5
6	110	{2, 3}	Rat 2 & 3 fed from bottle 6
7	111	{1, 2, 3}	Rat 1, 2 & 3 fed from bottle 7
8 (may be designated as zero)	000	{ }	No rat is fed from bottle 8 (or 0 if so designated)

It can be seen that rat 1 and 3 is fed from bottle 5, whose binary representation is 101, which indicates the position of three rats {3, 2, 1}, whether being fed from this bottle or not. The rat is being fed from a bottle if its positional bit is 1 otherwise not. Since the positional bit of rat 2 is 0, so rat 2 is not being fed from bottle 5. Now let's see, how to identify the poisoned bottle after the rat/rats are being fed from their respective bottle/bottles. It can be seen that rat 1 is fed from bottles 1, 3, 5 and 7.

Rat 2 is fed from bottles 2, 3, 6 and 7.

Rat 3 is fed from bottles 4, 5, 6 and 7.

Identify the rats which die within 12 hours.

Suppose only rat 2 dies, it indicates that bottle 2 is poisoned because only rat 2 is fed from this bottle. If rats 2 and 3 both dies, it indicates that bottle 6 is poisoned because, from this bottle, only rats 2 and 3 are fed. If no rat dies, then bottle 8 (0 if so designated) is poisoned as no rat was fed from this bottle. In this way, poisoned bottles can be identified out of $2^3 = 8$ bottles using only 3 rats. Let's now consider 120 bottles as given in the puzzle. Since $2^7 > 120$, so 7 rats are sufficient to identify poisoned bottles from 120 bottles of drink. The binary representation of every bottle will consist of 7 bits, which will uniquely define every bottle.

Table 4.24 shows some of the bottle numbers along with their 7-bit binary representation.

Table 4.24 .

Bottle numbers		Remarks
Decimal	Binary	
1	0000001	Rat 1 is fed from bottle 1
2	0000010	Rat 2 is fed from bottle 2
3	0000011	Rat 1 and 2 are fed from bottle 3
⋮	⋮	⋮
15	0001111	Rats 1, 2, 3 and 4 are fed from bottle 15
16	0010000	Rat 5 is fed from bottle 16
⋮	⋮	⋮
100	1100100	Rats 3, 6 and 7 are fed from bottle 100
101	1100101	Rats 1, 3, 6 and 7 are fed from bottle 101
⋮	⋮	⋮
119	1110111	Rats 1, 2, 3, 5, 6 and 7 are fed from bottle 119
120	1111000	Rats 4, 5, 6 and 7 are fed from bottle 120

It can be seen that rat 1 is fed from those bottles whose binary representation has a bit 1 in the unit's place. Similarly, rats 2, 3, 4 … are fed from those bottles whose binary representation has a bit 1 in the tens place, hundreds place and so on.

To identify the poisoned bottle, rats which are died after 12 hours are identified. Suppose only rat 2 dies which means the binary representation of the poisoned bottle must have a bit 1 in the tens place (2nd place from right) and 0 at all other places, so bottle number with binary representation as 0000010 i.e., bottle 2 is poisoned.

Similarly, if rat 1, 3, 6 and 7 dies, the binary representation of the poisoned bottle must be such that bit 1 occurs at 1st, 3rd, 6th and 7th place from the right and bit 0 at all other places. So, the bottle number with binary representation as 1100101 is poisoned, whose decimal equivalent is $2^6 + 2^5 + 0 + 0 + 2^2 + 0 + 1 = 101$.

In this way, the poisoned bottle can be identified out of 120 suspected bottles of drinks with the help of 7 rats.

87. **Black and White Balls**

Let's tabulate the given data in Table 4.25.

Table 4.25 .

Box	Number of black balls	Number of white balls
A	x	3x
B	6y	y
C	z	2z

Total number of black balls $= x + 6y + z = 50$ (i)

Total number of white balls $= 3x + y + 2z = 50$ (ii)

Eliminate x by multiplying Eq. (i) by 3 and then subtracting Eq. (ii)

$$17y + z = 100 \Rightarrow z = 100 - 17y$$
$$\Rightarrow y < 6$$ (iii)

From Eqs. (i) and (ii), eliminate z by multiplying Eq. (i) by 2 and then subtracting Eq. (ii)

$$-x + 11y = 50 \Rightarrow x = 11y - 50$$
$$\Rightarrow y > 4$$ (iv)

From Eqs. (iii) and (iv), we get y = 5, so

$$z = 100 - 17y = 15$$
$$x = 11y - 50 = 5$$

Tabulate the values of x, y and z in Table 4.26.

Table 4.26 .

Box	Number of black balls	Number of white balls
A	5	15
B	30	5
C	15	30
Total	50	50

88. Missing Page Numbers

Some of the important points in such puzzles are
  (i) The sum of page numbers on the two faces of a leaf is always odd.
  (ii) Each leaf starts with an odd number on the front page and an even number on the back page.

Sum of page numbers from 1 to 100

$$= \frac{100 \times 101}{2} = 5050$$

After tearing off some leaves, the sum of the remaining pages = 5005.
So, the sum of page numbers of torn leaves = 5050 − 5005 = 45.
Since the sum of page numbers of torn leaves is odd, so the number of torn leaves must be odd.

Let number of torn leaves = 1, so sequential page numbers of torn leaf will be 22 and 23 as 22 + 23 = 45.

This is not possible, as the first page (front page of the torn leaf) has to be odd because the numbering of pages starts from 1. Let's now assume the number of torn leaves = 3, so the number of pages of the torn leaf will be 6.

Let the first page of torn leaf = n, which is odd.

Sum of page numbers of three torn leaves

$$= n + (n+1) + (n+2) + (n+3) + (n+4) + (n+5)$$
$$= 6n + 15$$

Now 6n + 15 = 45 $\Rightarrow$ n = 5, which is odd.

Hence three leaves are torn with page numbers (5, 6), (7, 8) and (9, 10).

If we consider the number of torn leaves = 5 or more, then it will have 10 or more pages, whose sum will be 55 or more, which is not possible as the sum of page numbers of torn leaves is 45 only.

Comments: Another variation of this puzzle can be as follows:

If one leaf is torn from a book sequentially numbered from 1 to N and the sum of the remaining page numbers is 1600, find the page numbers of torn leaf and the total number of pages in the book i.e., N.

This can be solved as follows:

Let page numbers of the torn leaf be n and n + 1, where n is odd (i.e., the front page of the leaf).

The sum of the total page numbers of the book = $\frac{N(N+1)}{2}$

Sum of remaining page numbers = $\frac{N(N+1)}{2} - n - (n + 1) = 1600$

$$\Rightarrow N^2 + N = 4n + 3202 \tag{i}$$

The minimum and maximum values of n are 1 and N − 1 respectively. A range of N can be obtained using these limiting values of n.

For n = 1, $N^2 + N = 3206$

$$\Rightarrow N = \frac{-1 \pm \sqrt{1 + 4 \times 3206}}{2} = 56.12$$

For n = N − 1,

$$N^2 + N = 4(N - 1) + 3202$$
$$\Rightarrow N^2 - 3N - 3198 = 0$$
$$\Rightarrow N = \frac{3 \pm \sqrt{9 + 4 \times 3198}}{2} = 58.07$$

Since, N has to even, so N = 58.

From Eq. (i) 58 × 58 + 58 = 4n + 3202

$$\Rightarrow n = 55$$

So, the total number of pages in the book is 58. The page numbers of the torn leaf are 55 and 56.

89. **How Many Byes**

(i) If in a round, the number of players is even, then half the players are eliminated and the remaining half goes to the second round. If the number of players in any round is odd, say k, then one bye is given, so the number of players eliminated in that round is (k − 1)/2 and the remaining number of players i.e. (k + 1)/2 advances to the next round. So, if the number of players is 100, then in such a tournament.

$$100 \rightarrow 50 \rightarrow 25 \rightarrow 13 \rightarrow 7 \rightarrow 4 \rightarrow 2 \rightarrow 1$$

In general, if the number of players participating is n and the nearest power of 2 equal to or greater than n is 2^m, then m rounds are to be played in the tournament to decide the winner.
For n = 100, 100 < 2^7, so m = 7
So, seven rounds are to be played if hundred players are participating in the single-elimination tournament.

(ii) Since the number of players participating is not equal to a power of 2, some players need to be given bye's, which permit the player (who gets the bye) to go to the next round directly, without playing as there is no opponent.

For 100 players, the nearest power of 2 is equal to or greater than the number of players is 2^7 = 128.
The difference is 128 − 100 = 28.
If all the byes are given in the first round, so that in all subsequent rounds, the number of players is a power of 2, no byes are required in subsequent rounds. In such a case 28 byes are required in the first round. Since we are required to find the minimum number of byes and it is not compulsory to give all the byes in the first round itself, the number of bye's required to be given can be reduced as computed below:
The basic requirement is that in each round, the number of players must be even and if not, one bye is to be added in that round to make it even.
For 100 players, no bye is required in the first round and second round. In the third round, the number of remaining players is odd i.e., 25, so one bye is added to make it even i.e., 26. In the fourth round again, the number of remaining players is odd i.e., 13, so one bye is added to make it even i.e., 14. In the fifth round again, the number of remaining players is odd i.e., 7, so one bye is added to make it even i.e., 8. Since this is a power of 2, so no more byes are required

in any subsequent rounds as number of remaining players is even i.e., 4 in the sixth round and 2 in the seventh round, which decides the winner. So, the total number of byes required in this case is three i.e., in the 3rd, 4th and 5th rounds. Three byes are the minimum required for 100 players in a single-elimination tournament.

If n is the number of players and 2^k is the nearest power equal to or greater than n, then the number of byes required can be calculated as follows:

Let $m = 2^k - n$

Represent m in a binary system (base-2) and the number of 1's will give the number of byes required.

For $n = 100$, $n = 2^7 - 100 = 28$

$28_{10} = 11100_2$

Since there are three 1's, hence three byes are required.

90. Hourly Consumption of Lemon Juice

At a first glance, this puzzle appears to be complex, if one attempts to compute the proportion of lemon and water consumed each hour. However, the problem can be easily solved in the following two ways:

(i) Original bottle of 500 ml of pure lemon juice does not contain any water. Since in the end, the bottle is emptied, so total water consumed by the person will be equal to the total water added to the bottle.

$$= 50 + 100 + 150 + 200 + 250 + 300 + 350 + 400 + 450$$
$$= 2250 \, ml$$

(ii) Total water consumed by the person

= Total volume of liquid consumed by person − Volume of pure lemon juice available at start i.e. 500 ml

$$= (50 + 100 + 150 + \cdots + 450 + 500) - 500$$
$$= 2250 \, ml.$$

So, the person consumed 2250 ml of water and 500 ml of pure lemon juice.

91. Unique Four Digit Square

Let the four-digit perfect square is aabb $= x^2$,

$$aabb = 1000a + 100a + 10b + b$$
$$= 11(100a + b)$$

Since aabb is a perfect square, (100 a + b) must be divisible by 11.
100 a + b can be written as a0b. For a0b to be divisible by 11, a + b = 11

The last digit (unit's digit) of a perfect square can never be 2, 3, 7 or 8, so b cannot be 2, 3, 7 or 8.

Since a + b = 11, so b cannot be 0 or 1 as a and b are single-digit numbers. So, the possible values of b are 4, 5, 6 or 9.

If b = 5, its tens digit must be 2 and cannot be equal to b i.e., 5.

If b = 6, its tens digit must be odd and cannot be equal to b i.e., 6.

If b = 9, its tens digit must be even and cannot be equal to b i.e., 9.

So, the only possible value of b is 4.

Since a + b = 11, so a = 11 − 4 = 7

So, the required four-digit perfect square is $7744 = 88^2$.

92. The Age of Three Children

At a first glance, the information given appears insufficient to determine the ages. But on listing all possible combinations of ages of three children, whose product is 96, as given in Table 4.27, the solution appears feasible.

Table 4.27 .

Product	Possible ages of three children	A possible sum of ages
96	1, 1, 96	98
96	1, 2, 48	51
96	1, 3, 32	36
96	1, 4, 24	29
96	2, 2, 24	28
96	1, 6, 16	23
96	2, 3, 16	21
96	1, 8, 12	21
96	2, 4, 12	18
96	2, 6, 8	16
96	3, 4, 8	15
96	4, 4, 6	14

From Table 4.27, it can be seen that there are 12 triples of whole numbers (possible ages of three children) whose product is 96. The corresponding sum is also shown in Table 4.27.

If the sum of all 12 triples were different in all cases, it would have been possible for Suresh to determine the ages of the three children as he knew the house number which is the sum of the ages of three children.

But the sum of two combinations of three ages is the same i.e., 21. This is the reason Suresh was unable to find the ages of three children despite knowing the sum of ages of three children i.e., 21. So obviously house number is 21. This gives the two possibilities for ages of three children as 2, 3, 16 or 1, 8, 12.

The last clue given by Dinesh (which decides the ages of the three children) that "Your son is older than all my children" and also confirmation that Suresh will now know the ages decides the ages of the three children as follows:

If Suresh's son is older than 16, then both triples become possible as the ages of three children as it satisfies the condition that Suresh's son is older than all the children. In this case, neither Dinesh nor Suresh could have confirmed the solution to the puzzle. So, Suresh's son must be older than all children in one set of triples only and younger than at least one number in the other set. So, the age of Suresh's son must be greater than 12 but less than 16 years. So, the ages of three children will be 1, 8 and 12.

Comments: This puzzle has been appearing in literature for a long time as a census taker problem. Though the puzzle is appearing in different variations the concept remains the same as follows:

The product of three ages is normally specified in these puzzles. For example, the product specified is 36, 72, 1296, 2450 etc. This is interesting to know how these numbers are selected.

It is interesting to note that all such products must have exactly two triples with an equal sum. Some of the products with this property are given in Table 4.28. It may be an interesting exercise to find the treasure of such numbers.

Table 4.28 .

Product	Triple	Sum
36	2, 2, 9	13
	1, 6, 6	13
40	1, 5, 8	14
	2, 2, 10	14
72	2, 6, 6	14
	3, 3, 8	14
96	1, 8, 12	21
	2, 3, 16	21
1296	1, 18, 72	91
	2, 8, 81	91
2450	5, 10, 49	64
	7, 7, 50	64

93. **Pairs of Friendly Rectangles**

To avoid duplicates, let $a \geq b$, $c \geq d$ and $b \geq d$. The area and perimeter of the first rectangle with sides (a, b) are ab and 2 (a + b) respectively.

Area and perimeter of second rectangle with sides (c, d) is cd and 2 (c + d) respectively.

So,

$$ab = 2(c + d) \tag{i}$$

$$cd = 2(a + b) \tag{ii}$$

$$\Rightarrow abcd = 4(c + d)(a + b) \tag{iii}$$

Since $b \le a$ and $d \le c$, so

$$4(c + d)(a + b) \le 4(c + c)(a + a)$$
$$\Rightarrow 4(c + d)(a + b) \le 16ac \tag{iv}$$

From Eqs. (iii) and (iv)

$$abcd \le 16ac$$
$$\Rightarrow bd \le 16$$

Since $d \le b$, so $d^2 \le 16 \Rightarrow d \le 4$
It is given that a, b, c and d are even numbers, so d can only be 2 or 4.
For d = 2, b = 2, 4, 6 or 8   as $bd \le 16$.
For d = 4, b = 4.
Once values of b and d are fixed, the values of a and c can easily be found by expressing a and c in terms of b and d from Eqs. (i) and (ii).
Eliminating c from (i) and (ii), we get

$$a = \frac{4b + 2d^2}{bd - 4} \tag{v}$$

Eliminating a from Eqs. (i) and (v), we get

$$c = \frac{4d + 2b^2}{bd - 4} \tag{vi}$$

By putting the values of b and d as obtained above, values of a and c can be obtained from Eqs. (v) and (vi) as follows:
For d = 4 and b = 4,

$$a = \frac{4 \times 4 + 2 \times 4^2}{4 \times 4 - 4} = 4$$
$$c = \frac{4 \times 4 + 2 \times 4^2}{4 \times 4 - 4} = 4$$

Since both a and c obtained above are also even, so this is a valid pair of friendly rectangles i.e. (4, 4), (4, 4).
For d = 2, and b = 2, a and c are not valid as the denominator gives zero
For d = 2 and b = 4, a = 6 and c = 10 (valid pair)
For d = 2 and b = 6, a = 3 and c = 10 (since a is odd so not a required pair)
For d = 2 and b = 8, a = 40/12 and c = 136/12 (not integer values)

So, we get two friendly pair of rectangles as
(4, 4), (4, 4) and (6, 4), (10, 2)

Comments: For a rectangle with integer sides as (4, 4), area and perimeter are the same i.e., 16 and hence other pairs with the same sides will be a friendly rectangle. There are two such cases, where the sides of a rectangle are such that the area is equal to the perimeter. Let sides of such rectangle be (a, b), so

$$ab = 2(a+b)$$
$$\Rightarrow (a-2)(b-2) = 4$$

Two pairs of factors of 4 are 1×4 and 2×2, so a = 6, b = 3 or vice versa and a = 4, b = 4. Two pairs of friendly rectangles that can be formed from the above two rectangles are (4, 4), (4, 4) and (6, 3), (6, 3).
If the condition of all even sides is removed in the given puzzle and only the positive integer sides condition is retained, then all possible values of a and c can be obtained from the possible values of b and d, where $bd \leq 16$, $d \leq 4$ and $bd > 4$ (as can be inferred from denominators of Eqs. (v) and (vi). Discarding non-integer values of a and c, we can find that there are only seven pairs of friendly rectangles with integer sides. These are
[(a, b), (c, d)] = [(4, 4), (4, 4)], [(6, 3), (6, 3)], [(6, 4), (10, 2)], [(10, 3), (13, 2)], [(10, 7), (34, 1)], [(13, 6), (38, 1)], [(22, 5), (54, 1)]

94. **When RIGHT Becomes LEFT**

Let the required integer is X, with n digits and the rightmost digit as R.
Let the multiplier is M = 8.
Let's assume n = 5, so let X = ABCDR where R is the rightmost digit of X.
Now ABCDR $\times$ 8 = RABCD (R becomes the leftmost digit on multiplying by 8).
From this relation, it can be seen that if R < 8, then ABCDR becomes a four-digit number so $R \geq 8$. Let's first take R = 8, with multiplier M = 8, number of digits of X as n, we get

$$R \times 10^{n-1} + (X-R)/10 = MX$$
$$(10M-1)X = R(10^n - 1)$$

Putting the values of R and M. we get

$$X = \frac{8(10^n - 1)}{79}$$

So, we have to find the value of n such that X is an integer. The minimum such value of n is 13, which gives X = 1012658227848.
Similarly, if R = 9, we get $X = \frac{9(10^n - 1)}{79}$. Again, for X to be an integer, n = 13 gives X = 1139240506329.

Since the smallest integer is required in the puzzle, so X = 1012658227848
If we are not able to find the value of n, we can proceed as follows:
Let the n-digit number $X = m_{n-1} \ldots m_3 m_2 m_1 R$
For R = 8,

$$
\begin{array}{r}
m_{n-1} m_{n-2} \ldots m_3 m_2 m_1 8 \\
\times 8 \\
\hline
8\, m_{n-1} m_{n-2} \ldots m_3 m_2 m_1
\end{array}
$$

Now it is easy to get values of digits m_1, m_2, m_3 … as follows:
$8 \times 8 = 64$, so $m_1 = 4$, with 6 as carry over. So,

$$
\begin{array}{r}
m_{n-1} m_{n-2} \ldots m_3 m_2 4\ 8 \\
\times 8 \\
\hline
8\, m_{n-1} m_{n-2} \ldots m_3 m_2 4 \\
6
\end{array}
$$

Now $8 \times 4 = 32$, add carry over 6, we get 38, so $m_2 = 8$ with carry over as 3.
So,

$$
\begin{array}{r}
m_{n-1} m_{n-2} \ldots m_3 8\ 4\ 8 \\
\times 8 \\
\hline
8\, m_{n-1} m_{n-2} \ldots m_3 8\ 4 \\
3
\end{array}
$$

Continue the process in this way, till you find the repetition of the periodic sequence i.e., 1012658227848, which is the value of desired number X, as $1012658227848 \times 8 = 8101265822784$

Comments: If the multiplier is same as the rightmost digit R, then for R = 2 to 9, the values of X are:
R = 2, X = 105263157894736842 (18 digit),
R = 3, X = 10344827586206896551172413793 (28 digit),
R = 4, X = 102564 (6 digit),
R = 5, X = 142857 (6 digit),
R = 6,
X = 1016949152542372881355932203389830508474576271186440677966 (58 digit),
R = 7, X = 1014492753623188405797 (22 digit),
R = 8, X = 1012658227848 (13 digit),
R = 9, X = 10112359550561797752808988764044943820224719 (44 digit).

95. **Seventh Root**

(i) The unit's digit of the given number is 1, so the unit's digit of the seventh root must also be 1.

(ii) The number of digits in the given number = 14, so the number of digits in its seventh root is 2.

Since unit's digits of seventh root is 1, so only the remaining digit remains to be found out.

(iii) The digital root of given number is

$$= 5 + 1 + 6 + 7 + 6 + 1 + 0 + 1 + 9 + 3 + 5 + 7 + 3 + 1 = 55 \text{ and}$$
$$5 + 5 = 10, 1 + 0 = 1$$

It is obvious that, if the digital root of a number is 1, the digital root of any power of it must also be 1.

So, the second digit of seventh root must be 9, to make the digital root of 91 = 9 + 1 = 10, 1 + 0 = 1.

Hence the value of seventh root of the given number is 91.

Comments: Digital root of a number is obtained by taking the sum of the digits of the number repeatedly until single digit is obtained. For example, digital root of 397 is 1 as 3 + 9 + 7 = 19, 1 + 9 = 10 and 1 + 0 = 1.

96. **Motion on a Straight Road**

The ratio of initial speeds of vehicles X and Y = 3 : 2

Let's divide the 150 km route AB into five segments i.e., AC, CD, DE, EF and FB each measuring 30 km, so that vehicle with a speed of 90 km/h will cover 3 segments in the same time i.e., 1 hour. as the vehicle with 60 km/h will cover 2 segments. The details are shown in Fig. 4.47. So, when vehicle X reaches point E, vehicle Y starting from B will also reach point E, where both vehicles meet. At point E, both vehicles interchange their speed and reverse their direction of motion. So, from E, vehicle X travels towards A with 60 km/h speed and vehicle Y travels towards B with 90 km/h. Vehicle Y with 90 km/h will reach point B and reverse direction without changing speed, so will reach point F in one hour. Vehicle X with 60 km/h speed will reach point C in one hour. The vehicles continue to run at their speed in the direction of their motion unless the endpoint is encountered or they meet at some point. Vehicle X from C will reach endpoint A and return to point C, where it will meet vehicle Y, which starting from point F, also reaches point C. So, point C will be the second point of meeting.

Similarly, we can find the third meeting point i.e., at B as shown in Fig. 4.47.

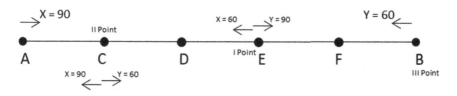

Fig. 4.47 .

The details are summarized in Table 4.29.

Table 4.29 .

Point	Direction of travel of vehicle X	Direction of travel of vehicle Y	Speed of X km/h	Speed of Y km/h	Distance travelled by	
					Vehicle X km	Vehicle Y km
Starting points A & B	Toward B	Towards A	90	60	-	-
First meeting point E	Towards A	Towards B	60	90	90	60
Second meeting point C	Towards A	Towards B	90	60	120	180
Third meeting point B	–	–	–	–	180	120
Total					**390 km**	**360 km**

So total distance travelled by vehicle X = 90 + 120 + 180 = 390 km and total distance travelled by vehicle Y = 60 + 180 + 120 = 360 km.

97. **Unlucky Thirteen**

Let's consider a number n with d digits.
The maximum value of $n = 10^d - 1$ and the minimum value of $n = 10^{d-1}$
For d = 1,

$$\text{Maximum value of } n = 9$$

$$\text{Minimum value of } n = 1$$

and sum of digits of n = n (being a single digit number).
Since n < 13 n, so d = 1 is not possible.
For d = 2,

$$\text{Maximum value of } n = 99$$
$$\text{Minimum value of } n = 10$$

If n is denoted by ab, then sum of the digits = a + b
So, $(a + b) \times 13 = 10a + b$
$\Rightarrow 3a + 12b = 0$
Since a and b are both positive integers, hence $3a + 12b$ is greater than 0, so
d = 2 is not possible.
For d = 3

$$\text{Maximum value of } n = 999$$
$$\text{Minimum value of } n = 100$$

If n is denoted by abc, then sum of the digits = a + b + c
So, $(a + b + c) \times 13 = 100a + 10b + c$

$$\Rightarrow 87a = 3b + 12c$$
$$\Rightarrow 29a = b + 4c$$

Since maximum value of a, b or c can be 9 only, so maximum value of 29a can
be $b + 4c = 9 + 4 \times 9 = 45$.
Hence the value of a cannot be more than 1. Also, the value of a cannot be zero,
as it will lead n to be a two-digit number, so a = 1.
Hence $29 \times 1 = b + 4c$
Since 29 is odd and 4c is even, so b must be odd.
For b = 1, c = 7 (possible)
For b = 3, c is not integer (Not possible)
For b = 5, c = 6 (possible)
For b = 7, c is not integer (Not possible)
For b = 9, c = 5 (possible)
Hence for three-digit numbers, there are three solutions i.e., 117, 156 and 195,
which are equal to 13 times the sum of their digits.
Now let's check for d > 3.
For d = 4, maximum value of n = 9999, maximum value of n = 1000.

$$\text{Maximum sum of digits} = 9 \times d = 36$$

Since $n \geq 10^{d-1}$ and maximum sum of digits = 9d, it can be seen that

$$10^{d-1} > 13 \times (9d) \text{ for all values of } d > 3.$$

Hence no number with four or more digits is possible.
Hence the required numbers are 117, 156 and 195 only.

98. **Doubly True Alphametic**

$$
\begin{array}{r}
\text{FORTY} \\
+\,\text{TEN} \\
+\,\text{TEN} \\
\hline
\text{SIXTY}
\end{array}
$$

(i) TY + EN + EN on dividing by 100 must leave a remainder of TY, hence EN = 50, So E = 5, N = 0.

(ii) S = F + 1, as the largest possible carry from the fourth column can be 1 only and S has to be different than F.

(iii) 10 + I = O + 1 or 10 + I = O + 2 (As carry over from third column can be 1 or 2).

If 10 + I = O + 1 then O = 9 and I = 0 (Not possible as N = 0).

Hence 10 + I = O + 2, So I = 1 and O = 9.

So, from third column 1 + R + T + T = 20 + X (Carry over is 2 from third to fourth column)

$$\Rightarrow R + 2\,T = 19 + X$$

Since digits 0, 1, 5 and 9 are already used by N, I, E and O respectively, we are left with 2, 3, 4, 6, 7 and 8.

So, R and T can only be 6, 7 or 8.

Since S = F + 1, so if F = 2, S = 3,

$$\text{if } F = 3, S = 4 \text{ hence } X \neq 3$$

So, X can only be 2 or 4

If X = 2, R + 2T = 21

If X = 4, R + 2T = 23, So, R must be odd.

Hence R = 7, from R + 2T = 21 ⇒ T = 7 (Not possible as R = 7)

Hence R = 7, from R + 2T = 23 ⇒ T = 8 and X = 4.

So, F = 2, S = 3 and the remaining digit is 6 which is equal to remaining letter Y. The final solution is

$$
\begin{array}{r}
29786 \\
+\,850 \\
+\,850 \\
\hline
31486
\end{array}
$$

99. **Tickets for the New Railway Station**

Since every station sells tickets to every other station, so each of the N railway stations will sell tickets to $N - 1$ different station. Hence originally $N \times (N - 1)$ different kinds of tickets were printed.

After the addition of M stations total station becomes $N + M$, which will sell tickets to $(N + M) - 1$ different station, so finally $(N + M) \times (N + M - 1)$ different kinds of tickets had to be printed.

So, number of additional kind of tickets to be printed (due to addition of M new stations)

$$= (N+M) \times (N+M-1) - N \times (N-1) = 62$$
$$\Rightarrow NM + NM + M^2 - M = 62$$
$$\Rightarrow M(2N+M-1) = 62$$

Since M and N are positive integers, M must be a factor of 62, so the value of M can be 1, 2, 31 or 62.

If M = 31 or 62, N becomes negative, so not admissible.

If M = 1, N = 31

If M = 2, N = 15

Both values of M i.e., 1 and 2 gives an admissible solution. However, on reading the puzzle carefully, it can be noted that "some new stations" were added, which implied that more than one station was added. In that case M = 2 and N = 15 is the correct solution.

So, the original number of stations = 15

Number of newly added stations = 2

100. **Egg Dropping**

Let egg safe floor be defined as the highest floor from which if an egg is dropped, it survives.

Suppose an egg is dropped from the 40th floor and it breaks. This gives us information that the egg safe floor must be below the 40th floor. Now if we drop the remaining egg from any floor from 39th to 2nd and if it breaks, we cannot be sure of egg safe floor so the solution cannot be obtained. So, the only option is to start from the 1st floor with the second egg. If the egg breaks by dropping it from the 1st floor, there is no egg safe floor but if it survives, then go to the 2nd floor and so on. If it breaks at say 7th floor, then the egg safe floor is the 6th floor. This has taken 1 + 7 = 8 droppings (trials). If the second egg breaks at say 39th floor, then the egg safe floor is the 38th floor, which has taken 1 + 39 = 40 droppings (trials).

From the above, it is clear that if only one egg is available, we have to test each floor, starting from the 1st floor for arriving at a solution. This is one egg solution. Now let's analyse the problem for two eggs only.

As explained earlier, if the first egg dropped from the 40th floor breaks, we start dropping the second egg from the 1st floor. In the worst-case scenario, we may have to go up to the 39th floor, so total drops (trials) in the worst-case scenario are 1 + 39 = 40.

Instead of starting from the middle floor of the building (i.e., the 40th floor), let's start from the 10th floor. If the first egg survives after dropping from the 10th floor, go to the 20th floor, then the 30th floor etc. up to the 80th floor and suppose the egg breaks at the 80th floor, then with the second egg start at the 71st floor and test each floor up to 79th floor (as egg safe floor lies between 71st to 79th floor), so the total number of drops = 8 + 9 = 17.

If the first egg breaks at the 10th floor, then with the second egg, start at the 1st floor and test each floor up to the 9th floor (as egg safe floor lies between the 1st to 9th floor), so the total number of drops = 1 + 9 = 10. So, in a worst-case scenario, there are 17 drops (trials). Instead of starting with the 10th floor, if we start with the 20th floor and move to the 40th, 60th and finally to the 80th floor, then the number of drops in the worst-case scenario = 4 + 19 = 23.

Let's now examine the case, where a limit is imposed on the maximum number of drops, say 10. Then how many floors can be covered optimally by two eggs.

Start with the 10th floor, if it breaks, with the second egg, start at the 1st floor and test each floor up to the 9th floor, so a total of 10 drops is sufficient. If the first egg survives after dropping from the 10th floor,

(i) for the second drop start from the 20th floor, if the egg breaks, then the second egg is tried from the 11th to 19th floor, so the total number of drops required in the worst-case scenario = 1 (10th floor drop) + 1 (20th floor drop) + 9 (11th to 19th floor drop).

Since the maximum limit of drops given is 10, so this option is ruled out.

(ii) since the first drop is already used at the 10th floor, so we are left with 9 drops only i.e., we can move only 9 floors. Hence the second drop is tried at the 19th floor instead of the 20th floor, if it breaks after dropping from the 19th floor, the second egg is tried from the 11th to 18th floor, so total drops become 1 + 1 + 8 = 10.

If the egg survives after the second drop from the 19th floor, we are left with 8 drops, so the next drop to be tried shall be (19 + 8) = 27th floor.

Keeping in view the limit of 10 drops, proceed in the same way as described above. The floors selected for first egg drop are summarized below:

$$10$$
$$10 + 9 = 19$$
$$10 + 9 + 8 = 27$$
$$10 + 9 + 8 + 7 = 34$$
$$10 + 9 + 8 + 7 + 6 = 40$$
$$10 + 9 + 8 + 7 + 6 + 5 = 45$$
$$10 + 9 + 8 + 7 + 6 + 5 + 4 = 49$$
$$10 + 9 + 8 + 7 + 6 + 5 + 4 + 3 = 52$$
$$10 + 9 + 8 + 7 + 6 + 5 + 4 + 3 + 2 = 54$$
$$10 + 9 + 8 + 7 + 6 + 5 + 4 + 3 + 2 + 1 = 55$$

So, if the number of floors is 55, then 10 drops are optimal.
Now let's find the drops required for 80 floors. Let number of drops required in worst case scenario are n.
As explained above, with n drops, we can cover

$$n + (n - 1) + (n - 2) + \cdots + 3 + 2 + 1 \text{ floors}$$
$$\Rightarrow 1 + 2 + 3 + \cdots + (n - 2) + (n - 1) + n \geq 80$$
$$\Rightarrow \frac{n(n + 1)}{2} \geq 80$$

Solving the above $n^2 + n - 160 = 0$

$$\Rightarrow n = \frac{-1 \pm \sqrt{1 + 640}}{2} = 12.16$$

So, 13 drops will be required for 80 floors in the worst-case scenario.
Check: $\frac{13 \times 14}{2} = 91 \geq 80$
In fact, with 13 drops, 91 floors can be covered.
With 12 drops, $\frac{12 \times 13}{2} = 78$ floors can be covered.
So, for 80 storey building, a minimum of 13 drops are required as explained below:
Drop the first egg from the 13th floor, if it breaks, then the second egg is to be tried from the 1st floor up to the 12th floor, so total drops = 1 + 12 = 13 in a worst-case scenario.
If the egg survives the first drop from the 13th floor, the second drop must be tried from the 25th floor (i.e., N − 1 floor above the Nth floor i.e., 13 − 1 = 12 floor above the 13th floor).
If it breaks, the second drop is tried from the 14th to 24th floor, so total drops are = 1 (13th floor) + 1 (25th floor) + 11 (14th to 24th floor) = 13.
So, 13 drops are sufficient in each case for an 80-floor building.

Comments: Suppose we remove the restriction of two eggs and have as many eggs as we require, then we can proceed as follows:

First drop the egg from the 40th floor, if it survives then the egg safe floor must be from the 40th floor to the 80th floor but if it breaks, then the egg safe floor must be from the 1st floor to the 39th floor. Proceeding in this way and selecting the drop at the middle by dividing the remaining possible floors in half. So, the drops (trials) required for 80 floors are 40, 20, 10, 5, 3, 2 and 1 (assuming that egg breaks in every drop).

The number of drops required is 7. In fact, with 7 droppings (trials), we can cover any building up to $2^7 = 128$ floors.

The number of drops in the worst-case scenario for 2 eggs for the given number of floors is tabulated in Table 4.30.

Table 4.30 .

Number of floors	1	2	3	5	10	20	50	100	200
Number of drops	1	2	2	3	4	6	10	14	20

The puzzle can be generalized to find the minimum number of drops (trials) required for m − storey building with n eggs.

The maximum number of floors which can be tested with 1 egg, 2 eggs or 3 eggs and with a given number of drops in the worst-case scenario is given in Table 4.31.

Table 4.31 .

Number of drops	Maximum number of floors which can be tested with		
	1 egg	2 eggs	3 eggs
1	1	1	1
2	2	3	3
3	3	6	7
4	4	10	14
5	5	15	25
6	6	21	41
7	7	28	63
8	8	36	92
9	9	45	129
10	10	55	175
11	11	66	231
12	12	78	298
13	13	91	377
14	14	105	469
15	15	120	575

It is interesting to note that 1 egg column is a sequence of natural numbers, 2 eggs column is the sequence of triangular numbers i.e., $^{n}C_1 + {}^{n}C_2 = \frac{n(n+1)}{2}$. Three eggs column is a 3-dimensional analogue of centered polygonal numbers given by

$$^{n}C_1 + {}^{n}C_2 + {}^{n}C_1 = \frac{n\,(n^2+5)}{6}$$

101. How Many Squares

No. of squares which can be easily visualized or located are (Figs. 4.48 and 4.49):

$$1 \times 1 \text{ squares} = 16(4 \times 4)$$
$$2 \times 2 \text{ squares} = 9(3 \times 3)$$
$$3 \times 3 \text{ squares} = 4(2 \times 2)$$
$$4 \times 4 \text{ squares} = 1(1 \times 1)$$

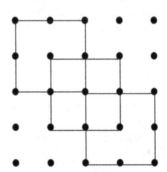

Fig. 4.48 Example of 2 × 2 squares. Total of 9 squares

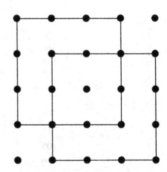

Fig. 4.49 Example of 3 × 3 squares. Total of 4 squares

So, the total number of squares of integer side = 1 + 4 + 9 + 16 = **30**
Let's consider the squares not so easily visible of type ♦ (Figs. 4.50, 4.51, 4.52, 4.53 and 4.54).

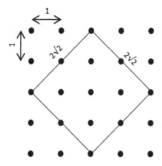

Fig. 4.50 One square of size $2\sqrt{2} \times 2\sqrt{2}$

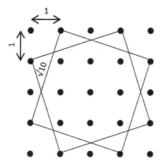

Fig. 4.51 Two squares of size $\sqrt{10} \times \sqrt{10}$

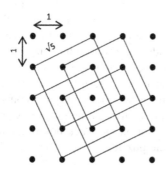

Fig. 4.52 Four squares of size $\sqrt{5} \times \sqrt{5}$

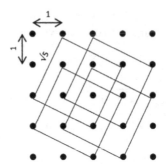

Fig. 4.53 Four squares of size $\sqrt{5} \times \sqrt{5}$

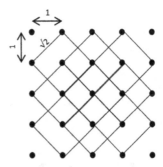

Fig. 4.54 Nine squares of size $\sqrt{2} \times \sqrt{2}$

So total number of squares with dots as corners $= 30 + 1 + 2 + 4 + 4 + 9 = 50$.

102. Identify Gold Coin Box

It is important to note that out of three statements, at least one is true and at least one is false. If the gold coins are in the black box, then as per the statement written on three boxes, all the three statements become true, which is contrary to the conditions given in the puzzle.
If the gold coins are in the white box, then as per the statement written on three boxes, all the three statements become false, which is again contrary to the conditions given in the puzzle. Hence the gold coins must be in the green box. In that case, statements written on the green box and white box become true and the statement written on the black box becomes false. This satisfied the conditions given in the puzzle.

103. Taxi Charges from Jaipur to Delhi

The distance given between two points = 100 km.

These two points are located on the route from Jaipur to Delhi, where an enquiry has been made from the driver regarding the distance covered/are to be covered, as shown in Fig. 4.55.

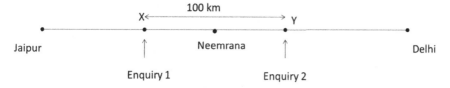

Fig. 4.55 .

Let the distance from X to Neemrana = a,
So, the distance from Jaipur to X = 2a
Let the distance from Neemrana to Y = b,
So, the distance from Y to Delhi = 2b
Distance from X to Y = a + b = 100 km,
So, the distance from Jaipur to Delhi

$$= 2a + a + b + 2b$$
$$= 3(a + b)$$
$$= 3 \times 100 = 300 \text{ km.}$$

The amount to be paid for taxi charges

$$= 300 \times 10 = 3000 \text{ rupees}$$

It may be noted that the time of starting from Jaipur and the time interval from Jaipur to the point, where the first enquiry is made are irrelevant and are not needed to solve the puzzle.

104. Cheryl's Birthday

In such types of puzzles, the conversation is sequentially examined statement by statement and possible options are eliminated based on the statements.
Albert has been told the month, so Albert is given May, June, July or August as the possible month of birth of Cheryl.
Bernard has been told the date, so Bernard is given 14, 15, 16, 17, 18 or 19 as the possible day of birth of Cheryl.
Now let's examine the conversation statement by statement.
(i)   Albert: "I don't know, when Cheryl's birthday is but I know that Bernard does not know too"
      Albert's statement that "I don't know when is Cheryl's birthday" is obvious as Albert only knows the month and every month has more than one option for the day of the birth, which is not known to Albert.

Albert's next part of the statement, "I know that Bernard does not know too" is possible if Bernard does not have a number that appears only once in the ten possible days given by Cheryl. Since numbers 18 and 19 appear only once and if Bernard is given one of these two numbers, then he will be able to deduce Cheryl's birthday because there is only one corresponding month for days 18 as well as 19.

Albert can be sure that Bernard does not know Cheryl's birthday implying that Bernard does not have 18 or 19, if Albert is given a month that does not have options of 18 or 19. This implied that Albert cannot have May or June as months, as these two months have options of 18 and 19 (i.e., June 18 and May 19). Hence Albert is given July or August as a month.

So, after eliminating five options, the remaining options are July 14, July 16, August 14, August 15 and August 17.

(ii) Bernard: "At first, I did not know when Cheryl's birthday is, but I know now."

This statement confirms that Bernard does not have 18 or 19 a day. But now Bernard knows when Cheryl's birthday is. After Albert's statement, he can deduce as explained above that Albert is given either July or August.

If Bernard is given 14, then he cannot deduce Cheryl's birthday as there are two month options with 14 as the day i.e. July 14 and August 14. So, Bernard is given 15, 16 or 17 and based on this, he can deduce Cheryl's birthday. Now the remaining options are July 16, August 15 and August 17.

(iii) Albert: "Now I also know when Cheryl's birthday is?"

The options available now are August 15, July 16 and August 17. For Albert to be certain now, he must have a month with only one option for the day i.e., July, as August has two options for the day i.e., 15 and 17, so Albert deduces July 16 as Cheryl's birthday.

Comments: This puzzle is talked about a lot in the literature. The origin of this puzzle dates back to 2006. Later, this puzzle is said to have been set for the 2015 Singapore and Asian School Math Olympiad contest based on similar entries available earlier.

When this puzzle was posted online in 2015, this caught a lot of public attention and found a place in various forums including newspapers and journals. When solutions were posted, a controversy was found between July 16 and August 17 as the correct answer depending upon the interpretation of the question. Later Singapore and Asian School Math Olympiad clarified the question that the natural interpretation of the question is "Albert deduces that Bernard does not know Cheryl's birthday" nor that he knew this information in advance whether being told or otherwise. With this clarification, they announced that the correct answer is July 16 and not August 17.

105. Twelve Dots and Connecting Lines

Before attempting a solution to the 12 dots problem, let us consider the famous 9 dots problem, where 9 dots are arranged in a square form divided into four squares as shown in Fig. 4.56.

Fig. 4.56 .

The problem is to find a minimum number of straight lines required to connect all nine dots, i.e., starting at a dot and ending at another dot, passing through each dot only once, and without lifting your pen from the paper.

While attempting this problem, due to psychological barriers, the imagination of most people limits to the boundaries within the square formed by nine dots and will use five straight lines to connect the nine dots simply as given in Fig. 4.57.

Fig. 4.57 .

But if we think beyond the boundaries of 9 dots, the solution can be obtained using only four straight lines as shown in Fig. 4.58.

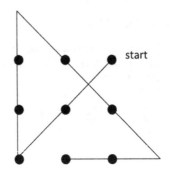

Fig. 4.58 .

There can be many other solutions which can be obtained by rotations/
reflections.

For a similar condition, if we attempt to solve the problem of 12 dots, then
(because of psychological barriers) we will come up with many solutions with
five straight lines, some of which are given in Fig. 4.59.

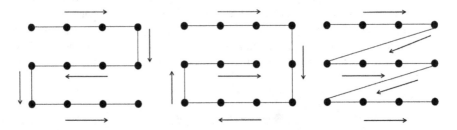

Fig. 4.59 .

But in the actual problem, one additional condition is given that the pen should
end up at the point where we started i.e., a closed solution is required. The
only closed solution leaving aside reflections/rotations satisfying the condition
can be obtained, if we think beyond the boundaries of 12 dots. The closed
solution for 12 dots requiring a minimum of five straight lines is given in
Fig. 4.60.

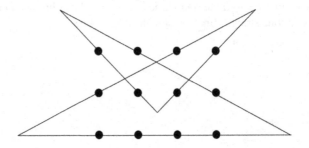

Fig. 4.60 .

Comments: We have seen that four straight lines are sufficient to connect 9 dots and five straight lines are sufficient to connect 12 dots. Also, it is not possible to connect 9 dots in less than four straight lines and 12 dots in less than five straight lines.

Let's see what happens in the case of 16 dots which are arranged in a square form and divided into nine smaller squares as shown in Fig. 4.61.

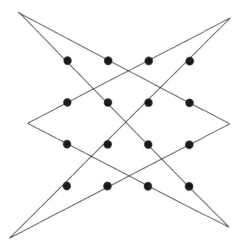

Fig. 4.61 .

For the similar conditions as given for 12 dots i.e., for a closed circuit, the pen should end up from where it started, a minimum of 6 straight lines are required and these are sufficient as given in Fig. 4.61. If this condition is removed then also a minimum of six straight lines are required to connect all 16 dots starting from a dot and ending on another dot. The solutions can be generated from the solution given in Fig. 4.61 as shown in Figs. 4.62 and 4.63.

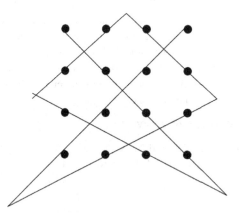

Fig. 4.62 .

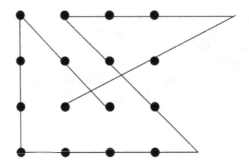

Fig. 4.63 .

It will be interesting to attempt all solutions for 16 dots which are more than 100.

106. Equalize Liquid Concentration

Let's denote the milk jar as M and the water jar as W.
Initially content of M is 100% milk and the content of W is 100% water.

(i) Transfer 100 ml of milk from jar M to jar W. Concentration of milk in jar M remains unchanged at 100%. The concentration of milk in jar W increases from 0% to say m%, where m < 100. Since jar M contains a higher concentration of milk than jar W, the transfer of liquid from jar M to jar W will not change the concentration of milk in jar M and will always have a higher concentration of milk as compared to jar W.

(ii) After stirring transfer 100 ml of the mixture from jar W to jar M.

The concentration of milk in jar W remains unchanged at m%, however, the concentration of milk in jar M reduces from 100% to say m1%, where m1 > m.

Since jar W contain a lower concentration of milk than jar M, the transfer of liquid from jar W to jar M will not change the concentration of milk in jar W and will always have a lower concentration of milk as compared to jar M.

If the process of transferring liquid from one jar to other continued, we will have one of the two cases mentioned above. So, jar M will always contain a mixture with a higher concentration of milk than in jar W.

The concentration of milk in both jars will never equalize unless we transfer the full content of one jar into the other jar.

107. Two Boats and Width of a River

Let boat at shore A be denoted by X_A and boat at shore B be denoted by X_B.
Let the width of the river is W. Refer to Fig. 4.64.

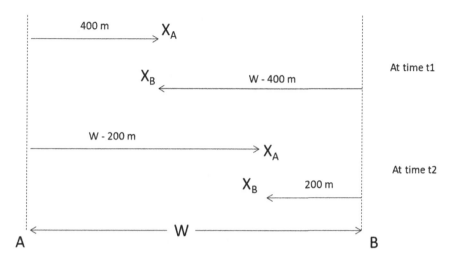

Fig. 4.64 .

The boats start at the same time, so when they cross the first time after time t_1,
$V_A t_1 = 400$ and $V_B t_1 = W - 400$,
where V_A and V_B are the speed of boats X_A and X_B respectively.

$$\text{So } \frac{V_A}{V_B} = \frac{400}{W - 400} \tag{i}$$

Let boat cross second time after time t_2, so
$V_B t_2 = 2W - 200$, $V_A t_2 = W + 200$

$$\frac{V_A}{V_B} = \frac{W + 200}{2W - 200} \tag{ii}$$

From Eqs. (i) and (ii), we get

$$\frac{400}{W - 400} = \frac{W + 200}{2W - 200} \Rightarrow W = 1000 \text{ m}$$

So, the width of the river is 1000 m.
This can also be easily obtained by comparing the ratios of distances on both trips. Since both travel at uniform speed, so the ratio of distances covered by both the boats on the first crossing $= \frac{400}{W - 400}$.
The ratio of total distance covered by both the boats on the second crossing $= \frac{W + 200}{2W - 200}$.

Comment: This puzzle can also be solved easily in the way detailed in Figs. 4.65, 4.66 and 4.67.

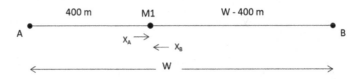

Fig. 4.65 .

Fig. 4.66 .

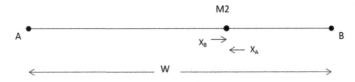

Fig. 4.67 .

The combined distance travelled by boats when they meet first time say at M1 = distance travelled by X_A + distance travelled by X_B

$$= AM1 + BM1$$
$$= W$$

The combined distance travelled by both boats when they reach opposite shores.

$$= \text{Distance travelled by } X_A + \text{Distance travelled by } X_B$$
$$= W + W = 2W.$$

The combined distance travelled by both boats when they meet second time, say at M2

$$= \text{Distance travelled by } X_A + \text{Distance travelled by } X_B$$
$$= W + BM2$$
$$+ W + AM2$$
$$= 2W + AM2 + BM2 = 3W.$$

Since both boats are travelling at a constant speed for the same period (At the meeting points M1 & M2), each boat must have travelled three times the distance travelled by each boat when they met the first time.
So, distance travelled by X_A from A to M2 (i.e., W + 200)

$$= 3 \times \text{distance travelled by } X_A \text{ from A to M1}$$

So, W + 200 = 3 × 400 = 1200 m

$$\Rightarrow W = 1000 \text{ m.}$$

In general, if the boats meet for the first time at a distance x metre from shore A and boats meet a second time at a distance y from shore B, then the width of the river = 3 x − y as already explained above by an actual example.

108. **First Day of a New Century**

There are 365 days in a common year and 366 days in a leap year. As per the Gregorian calendar, every year divisible by 4 is a leap year. However, every year divisible by 100 is not a leap year except when it is divisible by 400. So, years 2100, 2200 and 2300 are not leap years whereas 2000, 2400 and 2800 are leap years.

The present Century contains 24 leap years because 2100 is not a leap year. So, the number of days in the present Century = 36500 + 24 = 36524, which gives a remainder of 5, when divided by 7 (number of days in a week).

Since 1st January 2001 is Monday (Given), so 1st January 2101 will fall 5 days later than Monday i.e., Saturday. Similarly, we can deduce that 1st January 2201 will fall on Thursday and 1st January 2301 will fall on Tuesday. The details are summarized below:

1st January 2001 − Monday

1st January 2101 − Saturday (i.e. 5 days later)

1st January 2201 − Thursday (i.e. 5 days later)

1st January 2301 − Tuesday (i.e. 5 days later)

1st January 2401 − Monday (i.e. 6 days later, as 2400 is a leap year)

So, the first day of successive Centuries falls on Monday, Saturday, Thursday and Tuesday. Since every 400 years the cycle repeats, the first day of a Century will never fall on Wednesday, Friday or Sunday. So, people will never be able to celebrate New Century Day on Sunday.

Comments: In Gregorian calendar, the cycle repeats every 400 years as follows:

Number of days in 400 years $= 365 \times 400 +$ leap years

$$= 365 \times 400 + 97$$

$$= 146097, \text{which on dividing by 7 leaves no remainder.}$$

Hence cycle repeats every 400 years. There are 24 leap years in each of the three Centuries which are divisible by 100 but not 400 and the Century divisible by 400 will have 25 leap years.

So, total numbers of leap year in 400 years $= 24 \times 3 + 25 = 97$.

109. Largest Product from Given Sum

To maximize the product, avoid integer 1, as it will not increase the product but will increase the sum. If the product contains the factor 4, it can be replaced by 2×2, without affecting the sum and product. Any factor, $x > 4$ of the products can be replaced by $2 \times (x - 2)$, as for the same sum i.e., x, it will increase the product as $x < 2 \times (x - 2)$.

So, the factors in the product are either 2's or 3's, as 4 or more can be replaced by 2's and 3's or combination yielding higher product for the same sum. Now let's decide whether to have more 2's or more 3's.

Since $6 = 2 + 2 + 2 = 3 + 3$ and $3 \times 3 > 2 \times 2 \times 2$,

So, for the same sum i.e., 6, two threes result in a higher product than three twos. Hence every triplet of 2 shall be replaced by pair of 3's to maximize the product.

Similarly, $4 = 2 + 2 = 1 + 3$ and $2 \times 2 > 1 \times 3$,

So, for a given sum say N, the product can be maximized by successively dividing N by 3 i.e., powers of 3, till the remainder remains 4 or 2 which can be accounted for in terms of 2's. For maximum product, the given sum of 64 can be broken into twenty 3's (which gives the sum $= 60$) and two 2's (which gives the sum $= 4$) and their product $= 3^{20} \times 2^2$, is the largest possible product obtained from positive integers whose sum is 64.

Comments: It is a well-known fact that for two numbers whose sum is n, the largest possible product is given by $(n/2)^2$.

For example, if the sum of two numbers is 100, then the largest possible product of these two numbers is $(100/2)^2 = 2500$.

Similarly for three numbers, whose sum is n, the largest possible product is given by $(n/3)^3$.

If the sum is n and it is required to be broken into k parts (numbers), then to maximize the product, all those parts must be equal and the product is given by $(n/k)^k$. For example, if sum 8 is to be broken into five equal parts, then,

$$8 = 1.6 + 1.6 + 1.6 + 1.6 + 1.6 \text{ and product is } (8/5)^5.$$

It can be proved mathematically that to maximize the product, the value of $k = e$, where e is the base of the natural logarithm. The value of e is 2.71828...

To maximize the product of positive numbers with a given sum, each number must be as close to the value of e as possible.

Since $2 < e < 3$ and if only positive integers are considered, the sum is broken into numbers consisting of 2's and 3's, which are near to the value of e.

Since 3 is more closure to the value of e, the number is broken into 3's except the remainder number i.e., 2 or 4 as seen earlier.

To summarize, the procedure for finding a solution for the largest possible product which can be obtained from positive integers with sum as n is given below:

Divide n by 3 and let quotient is q and remainder be r. So, $n = 3q + r$,

If $r = 0$, then largest product $= 3^q$,

If $r = 1$, then largest product $= 3^{q-1} \times 2^2$,

If $r = 2$, then largest product $= 3^q \times 2$.

So, if $n = 64 = 3 \times 21 + 1$,

The largest product $= 3^{(21-1)} \times 2^2 = 3^{20} \times 2^2$

If $n = 100 = 3 \times 33 + 1$,

The largest product $= 3^{(33-1)} \times 2^2 = 3^{32} \times 2^2$.

110. Ten Digit Self-descriptive Number

Such numbers are termed as Self-descriptive numbers and can be found using various approaches. Let me describe two approaches.

(i) **First approach**: Let's describe the example given in the puzzle in a tabular form as given in Table 4.32.

Table 4.32 .

0	1	2	3	4
2	1	2	0	0

The numbers in the first row are called index numbers and corresponding columns are called the 0th column, 1st column, 2nd column etc. Each digit in the second row denotes the number of times the index number appears in the second row. Accordingly, the numbers in second row are 2 (denoting two 0's), 1 (denoting one 1), 2 (denoting two 2's), 0 (denoting zero 3's) and 0 (denoting zero 4's). So, 21200 is the 5-digit Self-descriptive number. Similarly let's find the 10-digit self-descriptive number, as shown in Table 4.33.

Table 4.33 .

0	1	2	3	4	5	6	7	8	9
9	0	0	0	0	0	0	0	0	0/1

Now we have to find suitable numbers for the second row.

Start with 9 as the first number and put it below the index number 0 in the 0th column. This indicates that there must be nine 0's, so all other nine digits must be zero. But since there is at least one 9, so 1 must be placed below index number 9 in the 9th column, hence all nine digits cannot be zero.

Now consider 8 as the first number. As discussed earlier, since there is at least one 8, so 1 must be placed in the 8th column below the index number 8, we get 8000000010. But now there is at least one 1, so 1 must occupy the position in the 1st column below the index number 1, we get 8100000010 but now zeros are only 7 instead of 8.

Next, consider 7 as the first number and proceed similarly, we can find that this also does not give a valid number.

Let's now consider 6 as the starting number. As explained earlier, the numbers generated sequentially are tabulated below in Table 4.34.

Table 4.34 .

Index number	0	1	2	3	4	5	6	7	8	9
Row 1	6	0	0	0	0	0	1	0	0	0
Row 2	6	1	0	0	0	0	1	0	0	0
Row 3	6	2	0	0	0	0	1	0	0	0
Row 4	6	2	1	0	0	0	1	0	0	0

We see that the final number in the 4th row i.e., 6210001000 satisfies the condition and is a valid 10-digit Self-descriptive number.

It is easy to verify that, if we take starting numbers like 5, 4, 3, 2 or 1, all turn out to be invalid, so it can be said that the only 10-digit Self-descriptive number is 6210001000.

(ii) **Second Approach**: For a 10-digit Self-descriptive number, since there are 10 digits in the number, the sum of all these 10 digits must be 10.

Partition the number 10 (i.e., sum) in all possible combinations of numbers from 1 to 9. There are 41 such partitions of the number 10.

We can start with the partition having the largest digit i.e., 9.

$9 + 1 = 10$, this gives 9000000001, where the first digit i.e., 9 denotes the number of 0's in the number. Since 9000000001 contain only eight 0's instead of 9, so this is not a valid Self-descriptive number.

Now consider,

$8 + 2 = 10$, this gives 8200000000, where the first digit 8 denotes the number of 0's and the second digit i.e., 2 denotes the number of 1's, which is not possible as zero's have already occupied all places.

Similarly, we can see that the following partitions do not result in a valid Self-descriptive number.

$$
\begin{aligned}
10 &= 8+1+1 \\
10 &= 7+3 \\
10 &= 7+2+1 \\
10 &= 7+1+1+1 \\
10 &= 6+4 \\
10 &= 6+3+1
\end{aligned}
$$

Let's now consider $10 = 6 + 2 + 1 + 1$. This gives 6210001000, which is a valid 10-digit Self descriptive number.

Comments: Some of the interesting observations for n-digit Self-descriptive numbers are:

a) The sum of the digits of the number must be n.
b) The general solution for n-digit Self-descriptive number, where $n > 6$ is given by

$$
X210_y1000
$$

where $X = n - 4$ and 0_y denotes zero repeated y times, where $y = n - 7$. For example, for $n = 9$, $X = 9 - 4 = 5$, $y = 9 - 7 = 2$, so the 9-digit Self descriptive number is 521001000. Similarly, 7-digit and 8-digit Self descriptive numbers can be computed as 3211000 and 42101000 respectively.

c) Two solutions for 4-digit Self-descriptive numbers are 1210 and 2020.

111. Upward Journey of a Ball

The initial drop of the ball $= 200$ m
First upward direction travel distance $= 200 \times \frac{1}{5}$ m
Then downward direction travel distance $= 200 \times \frac{1}{5}$ m (i.e., same as upward distance)
Second upward direction travel distance $= 200 \times \frac{1}{5} \times \frac{1}{5}$ m
Then downward distance $= 200 \times \frac{1}{5} \times \frac{1}{5}$ m
So, after the initial drop of the ball before hitting the ground, the ball moves upward and downwards continuously till it comes to rest.
After the initial drop of 200 m,

Upward direction distance covered by ball $=$ Downward distance

Total distance travelled by ball in upward direction

$$
\begin{aligned}
&= 200 \times \frac{1}{5} + 200 \times \frac{1}{5} \times \frac{1}{5} \cdots \\
&= \frac{200}{5} \left(1 + \frac{1}{5} + \frac{1}{5^2} + \frac{1}{5^3} + \cdots \right)
\end{aligned}
$$

Expression in bracket is an infinite geometric series whose sum is $= \frac{1}{1-\frac{1}{5}} = \frac{5}{4}$

So total upward direction distance $= \frac{200}{5} \times \frac{5}{4} = 50$ m

Total distance travelled by ball

$$= \text{Initial drop} + \text{Upward journey after first rebound}$$
$$+ \text{Downward journey after initial drop}$$
$$= 200 + 50 + 50 = 300 \text{ m.}$$

112. Mathematician on the Escalator

An escalator is a moving stair with a total number of fixed steps, which are difficult to count while they are moving. If you are moving with the escalator (i.e., in the direction of movement of the escalator), you would have to climb fewer steps, however, if you are moving against the escalator, you would have to climb more steps.

It is obvious that both of them are moving with the escalator and the faster man is taking more steps than his slower wife.

Let the speed of escalator $= v$ steps/second.

Let the speed of the mathematician's wife $= x$ steps/second, so the speed of the mathematician $= 2x$ steps/second.

When a mathematician climbs 28 steps at the speed of $2x$ steps/second, the time taken, say t is

$$t = \frac{28}{2x}$$

In this time t, escalator will move by $t \times v = \frac{28v}{2x}$

$$\text{So total number of steps in the escalator} = 28 + \frac{28v}{2x} \qquad \text{(i)}$$

When mathematician's wife climbs 21 steps at the speed of x steps/second, time taken $t_1 = \frac{21}{x}$

In this time t_1, escalator will move by $t_1 \times v = \frac{21v}{x}$

$$\text{So total number of steps in the escalator} = 21 + \frac{21v}{x} \qquad \text{(ii)}$$

Equating Eqs. (i) and (ii)

$$28 + \frac{28v}{2x} = 21 + \frac{21v}{x}$$

$$\Rightarrow \frac{14v}{2x} = 7 \Rightarrow v = x$$

So total number of steps $= 28 + \frac{28x}{2x} = 42.$

113. **Incorrectly Labelled Boxes**

Draw a coin from the box labelled GS. If the coin drawn is a gold coin, then the correct label of this box must be G containing all gold coins. Similarly, if the coin drawn is a silver coin, then the correct label of this box must be S, containing all silver coins. This box cannot contain a mix of gold and silver coins, as the label of box GS is incorrect.

Once incorrectly labelled box GS is correctly identified as G or S, then it is easy to correctly identify the other two incorrectly labelled boxes also.

If incorrectly labelled box GS is correctly identified as G, then incorrectly labelled box S will be GS, as it cannot be G which is already identified and incorrectly labelled box G will be S.

Similarly, if incorrectly labelled box GS is correctly identified as S, then incorrectly labelled box G will be GS, as it cannot be S, which is already identified and incorrectly labelled box S will be G.

So, only one trial is sufficient to identify the correct labels of all three boxes.

114. **How Long Professor Walked**

(i) Since they arrived 30 minutes earlier than usual, so 30 minutes were saved on driving by the son on the round trip i.e., from home to pick up point and back. That means 15 minutes were saved from home to pick up point (A) and 15 minutes from pickup point (A) to home. So, the son meets the professor 15 minutes earlier than the usual time of 18.00 hours i.e. 17.45 hours. Assume that the car goes from meeting point B to pick up point A, then it will take 15 minutes in going and 15 minutes in returning from pickup point to meeting point. So, if the car was to reach the pickup point at 18.00 hours, as usual, it will reach the meeting point 15 minutes earlier i.e., 17.45 hours. Therefore, it is 17.45 hours, when the professor meets his son at meeting point B.

(ii) Since the professor reached early at pickup point i.e., at 16.30 hours instead of usual 18.00 hours, so professor walked from 16.30 hours i.e., pickup point A to meeting point B at 17.45 hours, where he meets his son as explained in (a) above.

Hence professor walked for 17.45 − 16.30 = 75 minutes.

115. **Ratio of a Number to its Digit Sum**

Let the four-digit number is N = abcd, so
N = 1000a + 100 b + 10 c + d
Let the ratio of N to the sum of its digits is R, so

$$R = \frac{1000a + 100b + 10c + d}{a + b + c + d}$$

If b = c = d = 0, then R = 1000

If $b + c + d > 0$, then $a + b + c + d \geq a + 1$

Since a is the first digit of N and is in thousand's place,

So, $N < (a + 1) \times 1000$

$$\Rightarrow R = \frac{N}{a+b+c+d} < \frac{(a+1) \times 1000}{a+1}$$

$$\Rightarrow R < 1000$$

Hence the greatest value of the ratio of a four-digit number to its digit sum is 1000.

Comments: The greatest value of ratio of an n-digit number to its digit sum is equal to 10^{n-1}.

For $n = 1$, $R = 10^{n-1} = 1$

For $n = 2$, $R = 10$

For $n = 3$, $R = 100$ and so on.

It will be an interesting exercise to find the least value of the ratio i.e., R1.

For $n = 1$, least value of ratio i.e., $R1 = 1$.

For $n = 2$, $R1 = \frac{19}{10} = 1.9$

For $n = 3$, $R1 = \frac{199}{19} = 10.4737 \ldots$

For $n = 4$, $R1 = \frac{1099}{19} = 57.8421 \ldots$

Can you find the general expression?

116. Financial Assistance to Villagers

i) He can withdraw 1000 rupees only at a time, so distance covered by collector

$$= 6 \times 2 + (6 \times 2) \times 2 + (6 \times 3) \times 2 + \cdots + (6 \times 100) \times 2$$
$$= (6 \times 2)(1 + 2 + 3 + \cdots + 100)$$
$$= (12)\left(\frac{100 \times 101}{2}\right) = 60600 \text{ feet}$$

ii) In this case, he is permitted to withdraw 1000 rupees at a time till he distributes the money to first 50% of the persons. After that, he was permitted to withdraw 2000 rupees at a time till he completes distribution.

Distance covered in distributing the money to the first 50% of the persons.

$$= 6 \times 2 + (6 \times 2) \times 2 + \cdots + (6 \times 50) \times 2$$
$$= (6 \times 2)(1 + 2 + 3 + \cdots + 50)$$
$$= (12)\left(\frac{50 \times 51}{2}\right) = 15300 \text{ feet}.$$

Distance covered in distributing the money to balance 50% of the persons.

$$= (300 + 12) \times 2 + (300 + 2 \times 12) \times 2 + \cdots + (300 + 25 \times 12) \times 2$$
$$= 600 \times 25 + (12 \times 2)(1 + 2 + 3 + \cdots + 25)$$
$$= 15000 + 24 \times \frac{25 \times 26}{2}$$
$$= 15000 + 7800$$
$$= 22800 \text{ feet.}$$

So total distance covered = 15300 + 22800 = 38100 feet.

117. Gold Coins and Culprit Jeweller

If all gold coins were genuine, then total weight of gold coins would have been $100 \times 100 = 10000$ g.
However, the total weight of gold coins

$$= 90 \times 100 + 10 \times 99$$
$$= 9000 + 990 = 9990 \text{ grams}$$

To identify the culprit jeweller from J1, J2 ... J10, who supplied the lighter gold coins, proceed as follows:
Put the gold coins in stacks jeweller wise i.e., stack J1 contains 10 gold coins from jeweller J1, stack J2 contains 10 gold coins from jeweller J2 and so on. Now take one coin from stack J1, two coins from stack J2, three coins from stack J3 and so on till nine coins are taken from stack J9.
Total coins taken = 1 + 2 + 3 + 4 + 5 + 6 + 7 + 8 + 9 = 45.
If all 45 coins were genuine, their weight will be 4500 grams.
On single weighing, let the weight of these coins = (4500 − x) grams.
If x = 0, all coins from J1 to J9 are genuine, so J10 is the jeweller who supplied the lightweight coins.
If x = 1, then weight of 45 coins.

$$= 4500 - 1 = 4499 \text{ grams.}$$

So, there is only one lightweight coin out of 45 selected coins. Since one coin is taken from stack J1 hence stack J1 contains the lightweight coins and J1 is the culprit jeweller, who supplied lightweight coins. Similarly, if x = 2, 3, 4, 5, 6, 7, 8 or 9, then two, three, four, five, six, seven, eight or nine coins are lightweight coins respectively, which are taken from stack J2, J3 ... J9. So based on the identified stack, the corresponding jeweller can be identified, who supplied the lightweight coins.

118. Pricing of a Fruit Set

Let cost of one Apple is A, one Mango is M and one Banana is B. So

$$7A + M + 19B = 230 \qquad \text{(i)}$$

$$6A + M + 16B = 200 \qquad \text{(ii)}$$

Subtract Eq. (ii) from Eq. (i), we get

$A + 3B = 30$

$\Rightarrow A = 3(10 - B)$

So, B < 10 and A is a multiple of 3.

By putting any value of B from 1 to 9, value of A can be obtained.

For example,

For B = 1, A = 27

For B = 2, A = 24 and so on.

By putting any one of the above values of A and B in Eqs. (i) or (ii), we can find the value of M as follows:

$7 \times 27 + M + 19 \times 1 = 230$

$\Rightarrow M = 22$

So, A + M + B = 27 + 22 + 1 = 50.

Comments: This puzzle can also be solved in the following way.

Multiply Eq. (ii) by 6 and Eq. (i) by 5, we get

$$36A + 6M + 96B = 1200 \qquad \text{(iii)}$$

$$35A + 5M + 95B = 1150 \qquad \text{(iv)}$$

Subtract Eq. (iv) from Eq. (iii), we get

$$A + M + B = 50.$$

So, the combined cost of one apple, one mango and one banana are 50 rupees.

119. Perfect Powerful Number

Let the required smallest positive integer is n. For n/2 to be a perfect square, n must be even i.e., 2 must be a factor of n. For n/3 to be a cube, 3 must be a factor of n. Similarly for n/5 to be a fifth power and n/7 to be the seventh power, 5 and 7 must also be a factor of n respectively.

Let $n = 2^a . 3^b . 5^c . 7^d$

$$\text{Now } \frac{n}{2} = 2^{a-1} 3^b 5^c 7^d (\text{Perfect square}) \qquad \text{(i)}$$

$$\frac{n}{3} = 2^a 3^{b-1} 5^c 7^d (\text{Perfect cube}) \qquad \text{(ii)}$$

$$\frac{n}{5} = 2^a 3^b 5^{c-1} 7^d \text{(Perfect fifth power)} \tag{iii}$$

$$\frac{n}{7} = 2^a 3^b 5^c 7^{d-1} \text{(Perfect seventh power)} \tag{iv}$$

From Eq. (i), $(a - 1)$ must be even and from Eqs. (ii), (iii) and (iv), a must also be multiple of 3, 5 and 7 respectively, so smallest value of $a = 3 \times 5 \times 7 = 105$. This gives $a - 1 = 105 - 1 = 104$ i.e., even as required. Similarly, b must be even, $(b - 1)$ must be a multiple of 3 and b must be multiple of 5 and 7.

Let $b = 2 \times 5 \times 7 = 70$, so $b - 1 = 69$, which is a multiple of 3 as required, so smallest value of $b = 70$. From Eq. (i), c must be even, from Eq. (iii), $(c - 1)$ must be a multiple of 5 and c must be a multiple of 3 and 7 as per Eqs. (ii) and (iv) respectively.

Let $c = 2 \times 3 \times 7 = 42$ but $c - 1 = 42 - 1 = 41$ is not a multiple of 5. $c = 42 \times 2 = 84$ also does not satisfy the condition that $c-1$ must be a multiple of 5. So let $c = 42 \times 3 = 126$, so, $c - 1 = 125$, which is a multiple of 5 as required.

So smallest value of $c = 126$.

Similarly, d must be even, a multiple of 3 and 5, $d - 1$ must be a multiple of 7. Let $d = 2 \times 3 \times 5 = 30$ but $d - 1 = 30 - 1 = 29$ is not a multiple of 7. $d = 30 \times 2 = 60$ or $d = 30 \times 3 = 90$ also does not satisfy the condition that $d-1$ must be a multiple of 7, let $d = 30 \times 4 = 120$, so $d - 1 = 119$, which is a multiple of 7 as required. Thus, smallest value of $d = 120$. Hence smallest value of n to satisfy all given conditions is given by

$$\mathbf{n = 2^{105} 3^{70} 5^{126} 7^{120}}$$

Comments: Another variation of this puzzle can be as follows:

"Find the smallest positive integer that is one half of a perfect square, one-third of a perfect cube, one-fifth of a fifth power and one-seventh of a seventh power".

Let smallest such integer $= n$.

For the given conditions, 2, 3, 5 and 7 must be a factor of n.

Let $n = 2^a \, 3^b \, 5^c \, 7^d$

Since n is one half a perfect square, so 2n must be a perfect square, so

$$2^{a+1} 3^b 5^c 7^d \text{ (A perfect square)}$$

Similarly
$$2^a 3^{b+1} 5^c 7^d \text{ (A perfect cube)}$$
$$2^a 3^b 5^{c+1} 7^d \text{ (A perfect fifth power)}$$
$$2^a 3^b 5^c 7^{d+1} \text{ (A perfect seventh power)}$$

So, a must be a multiple of 3, 5, 7 and $(a + 1)$ must be even. Let $a = 3 \times 5 \times 7 = 105$, so $a + 1 = 105 + 1 = 106$, which is even as required, so $a = 105$.

Similarly, we can find that b = 140, c = 84 and d = 90 satisfying all given conditions.

Hence the smallest required number = $2^{105}\ 3^{140}\ 5^{84}\ 7^{90}$.

120. Minimum Number of Houses

Security post numbered as m divides equally the sum of house numbers on both sides of it, so

$$1 + 2 + 3 + \cdots + (m - 1) = (m + 1) + (m + 2) + \cdots + n \qquad \text{(i)}$$

Add $(1 + 2 + 3 + \cdots + m)$ on both sides of Eq. (i).

$$2(1 + 2 + \cdots + m - 1) + m = 1 + 2 + 3 + \cdots + m + m + 1 \cdots + n$$

$$\Rightarrow 2 \times \frac{(m - 1)m}{2} + m = \frac{n(n + 1)}{2}$$

$$\Rightarrow m^2 = \frac{n(n + 1)}{2} \qquad \text{(ii)}$$

$$\Rightarrow n^2 + n - 2m^2 = 0$$

So, $n = \frac{-1 \pm \sqrt{1 + 8m^2}}{2}$.

For n to be an integer, $1 + 8m^2$ must be a perfect square. Let $1 + 8m^2 = k^2$

For m = 0, k = 1,

m = 1, k = 3,

m = 6, k = 17.

For any house to exist, the security post number i.e., m must be greater than 1. So, the smallest value of m is 6, as obtained above. Substituting the value of m in Eq. (ii), we get n = 8. Since security post number m = 6, five houses are numbering 1, 2, 3, 4 and 5 on one side. Since n = 8, so remaining houses numbering 7 and 8 are on the other side. So, the minimum number of houses in the colony = 7.

Comments: Right-hand side expression of Eq. (ii) i.e. $\frac{n(n+1)}{2}$ is nth triangular number.

The triangular numbers are the numbers obtained as the sum of the first n natural numbers, so the nth triangular number is given by

$T_n = 1 + 2 + 3 + \cdots + n = \frac{n(n+1)}{2}$

For n = 1, 2, 3 ..., we get $T_1 = 1$, $T_2 = 3$, $T_3 = 6$ ... etc.

From Eq. (ii) $m^2 = \frac{n(n+1)}{2}$ defines triangular numbers which are perfect squares.

We have seen that for the nth triangular number to be a perfect square, $1 + 8m^2$ must be a perfect square say k^2. Hence $1 + 8\ m^2 = k^2$ as discussed above.

Now $1 + 8m^2 = k^2$ is a form of pell's equation $k^2 - 8m^2 = 1$ i.e., $x^2 - 8y^2 = 1$, the solution for which can be seen on line or from any standard text book.

121. Locker and Coins

Designate the lockers as A and B. Ask the person to multiply the number of coins in locker A by 2 and the number of coins in locker B by 3. If the sum of these two products is odd, then the gold coins will be in locker A and the silver coins in locker B. If the sum of these two products is even, then the gold coins will be in locker B and the silver coins in locker A.

For example, assume there are 6 coins in locker A and 7 coins in locker B. Since $6 \times 2 + 7 \times 3 = 33$ is an odd number, so gold coins (even number) will be in locker A.

If there are 7 coins in locker A and 6 coins in locker B, then gold coins will be in locker B as $7 \times 2 + 6 \times 3 = 32$ is an even number.

Comments: This can be proved as follows:
Let the number of coins in locker A be a and the number of coins in locker B is b. Let 2x be any even number and 2y + 1 be an odd number (x and y are integers).
Then $2xa + (2y + 1) b = 2(xa + yb) + b$
Since $2(xa + yb)$ is even, so if the sum of products is even, it implies that b is even which means gold coins will be in locker B. If the sum of products is odd, this implies that b is odd, so silver coins will be in locker B.

122. Sum of Digits of Numbers

Add zero at the start as it does not change the result. Now pair first with last number, second with second last number and so on. Find the sum of each pair as follows:

First pair	$0 + (10^n - 1) = 10^n - 1$
Second pair	$1 + (10^n - 2) = 10^n - 1$
Third pair	$2 + (10^n - 3) = 10^n - 1$

The Sum of digits of $10^n - 1$ is 9n. So, the sum of digits of each pair is 9n. We can see that there are $(10^n/2)$ pairs in total, so the sum of digits of all pairs $= 9n (10^n/2)$.

Comments: Suppose we are asked to find the sum of all the digits appearing in all three-digit numbers. There can be two ways to solve it.

a)   There are 900 three-digit numbers, so there are 900 unit's digits, 900 ten's digits and 900 hundred's digits.
   Now unit's digit can be 0 to 9 (total of 10 digits) and the sum of 0 to 9 is 45, so the sum of the unit's digits of all three-digit numbers $= 90 \times 45$.

Similarly sum of all ten's digits of all three-digit numbers = 90 × 45.
Since hundred's digits can only be 1 to 9 (Total 9 digits), so the sum of all
hundred's digits = 100 × 45.
So, sum of all digits appearing in all three-digit numbers = 90 × 45 + 90 ×
45 + 100 × 45

$$= 12600$$

b)   Calculate the sum of digits of all numbers up to three digits i.e., $9n \times 10^{n}/2 = 9 \times 3 \times (500) = 13500$

Now deduct the sum of digits of all numbers up to two digits i.e.

$$= 9 \times 2(50) = 900$$

So required sum = 13500 − 900 = 12600

123. Number of Diagonals in a Polygon

Consider an n-sided convex polygon.
For n = 3, we have a triangle, where we have 3 vertices and no diagonal.
For n = 4, we have a quadrilateral, which has 4 vertices and two diagonals.
The number of sides n = the Number of vertices in a polygon.
Number of vertices in n-sided polygon = n.
From each vertex, n − 3 diagonals can be drawn because from two adjacent
vertices, we get two adjacent sides which cannot be diagonals and a diagonal
cannot be drawn from a vertex to itself.
In a pentagon (n = 5), n − 3 = 5 − 3 = 2 diagonals can be drawn from any
vertex, say A as shown in Fig. 4.68.

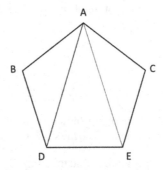

Fig. 4.68 .

Total number of vertices in an n-sided polygon = n
So total number of diagonals = n (n − 3)
It can be seen that the diagonal from vertex A to vertex E is the same as the
diagonal from vertex E to vertex A. So, in the formula given above, these

diagonals have been counted twice. So, accounting for this, total number of diagonals = n (n − 3)/2.

In the puzzle, total number of diagonals given = 65, so n (n − 3)/2 = 65

$$\Rightarrow n^2 - 3n - 130 = 0$$
$$\Rightarrow (n+10)(n-13) = 0$$

Ignoring the negative value of n, a polygon with 65 diagonals has 13 vertices and hence 13 sides, so this is a 13-sided polygon.

Comments: A diagonal is a line segment connecting two non-consecutive vertices in a polygon. The number of diagonals in an n-sided convex polygon that can be drawn from any vertex in a polygon is three less than the number of sides.

The total number of diagonals = Number of diagonals per vertex i.e. $\frac{(n-3) \times n}{2}$, as each diagonal is counted twice.

124. Peculiar Pages in a Book

Let's first find out the number of pages in the book.

There are N pages in the book. Looking at the total number of digits i.e., 792, it can be assumed that number of pages N in the book lies between 100 to 999. Since page numbers from 1 to 9 are single-digit numbers, page numbers from 10 to 99 are double-digit numbers and page numbers from 100 to N are three-digit numbers, so the total number of digits for N pages in the book is

$$9 + 2 \times (99 - 9) + 3 \times (N - 99) = 792$$
$$\Rightarrow N = 300$$

So, the number of pages in the book is 300.

Now we have to identify page numbers which contain the digit '1' at least once. There are many ways to do it. One of the simplest ways is as follows:

In this method, count the numbers which don't contain the digit '1' and then subtract them from the total numbers as given below:

(i)   For one-digit numbers 1 to 9, there are 8 numbers which don't contain the digit '1'.

(ii) For two-digit numbers, the unit's digit can have any number from 0 to 9 except 1 i.e., 9 cases and the ten's digit can have any number from 2 to 9 i.e., 8 cases. So total number of two-digit numbers which don't contain digit '1' = 8 × 9 = 72.

(iii) Since we have numbers only up to 300 and 300 don't contain the digit '1', it can be counted separately. For three-digit numbers from 100 to 299, unit's digit can have any number from 0 to 9 except 1 i.e., 9 cases, ten's digit can have any number from 0 to 9 except 1 i.e., 9 cases and hundred's digit can have only number 2 i.e., 1 case, so total cases = 9 × 9 × 1 = 81.

Adding 1 Case for 300 which also don't contain digit '1", number of three-digit numbers not containing digit '1'

$$= 81 + 1 = 82$$

Total numbers up to 300, not containing digit '1' = 8 + 72 + 82 = 162.
So, total numbers containing the digit '1' = 300 − 162 = 138.

Comments: The number of pages can also be determined as follows:
Every page number must have a unit's digit, so if we have N pages (i.e., from 1 to N), there are N digits in the unit's place.
Except for single-digit page numbers i.e., from 1 to 9, all other pages have a digit in the ten's place, so the total number of digits in ten's place = N − 9.
Except for the first 99 pages, all other pages have a digit in hundred's place, so the total number of digits in hundred's place = N − 99.
So total number of digits = N + (N − 9) + (N − 99) = 792
$\Rightarrow$ N = 300
Following comments are offered regarding the second part of the puzzle.
It is easy to find the solution, if the puzzle is like this, "How many numbers from 1 to 999 contain the digit 1".
It is easy to find that from 0 to 9, the numbers containing the digit '1' (or any other digit from 2 to 9) = $10^1 – 9^1$ = 1.
Similarly, the numbers containing the digit '1' (or any other digit from 2 to 9) from 0 to 99

$$= 10^2 - 9^2 = 19$$

From 0 to 999, the numbers containing the digit '1' (or any other digit from 2 to 9)

$$= 10^3 - 9^3 = 271 \text{ and so on.}$$

125. **Coin Distribution**

In such types of puzzles, it is important to look for patterns. Let's put the coins distributed house wise as shown in Table 4.35.

Table 4.35 .

Nth coin distributed to									
H1	H2	H3	H4	H5	H6	H5	H4	H3	H2
1	2	3	4	5	6	7	8	9	10
11	12	13	14	15	16	17	18	19	20
21	22	23	24	25	26	27	28	29	30
31	32	33	34	35	36	37	38	39	40
⋮	⋮	⋮	⋮	⋮	⋮	⋮	⋮	⋮	⋮
9991	9992	9993	9994	9995	9996	9997	9998	9999	10000

A pattern can be seen in Table 4.35. After distributing the coins in the manner described above, it can be seen that after giving the 10th coin to H2, the 11th coin is given to H1, so the cycle repeats after every 10 coins.

So, to know which house gets the 10000th coin, find the remainder obtained by dividing the total number of coins i.e., 10000 by the number of coins per cycle i.e., 10. Since the remainder is zero, the 10000th coin is given to H2 appearing after house H3 in Table 4.35. If the remainder is 1, 2, 3, 4, 5, 6, 7, 8 or 9, then the house H1, H2, H3, H4, H5, H6, H5, H5 and H3 respectively gets the last coin.

It can also be noted that in each cycle of 10 coins, houses H1 and H6 get one coin each, whereas houses H2, H3, H4 and H5 get two coins each. For 10,000 coins, there will be 1000 cycles of 10 coins each, so the number of coins received by houses H1, H2, H3, H4, H5 and H6 will be 1000, 2000, 2000, 2000, 2000 and 1000 respectively.

Comments: The puzzle can be modified to n number of houses with m number of coins to be distributed. It can be seen that in such a case, the cycle will repeat after $2(n-1)$ coins. To find the house number, which gets the last coin i.e., mth coin, divide m by $2(n-1)$ to obtain the remainder, say r. If $r = 1, 2, 3, 4 \ldots 2(n-1) - 1$ or 0, the mth coin will be received by the house $H_1, H_2, H_3, H_4, \ldots H_{n-1}, H_n, H_{n-1} \ldots H_3$ and H_2 respectively. The number of coins received by H_1 and H_n is 1 per cycle, whereas the number of coins received by all other houses is 2 per cycle.

Printed in the United States
by Baker & Taylor Publisher Services